Heinz Schumann and David Green

Discovering Geometry with a Computer

– using Cabri-Géomètre

Chartwell-Bratt   Studentlitteratur

British Library Catalogouing in Publication Data
A catalogue record for this book is available from the British Library

⚠

All rights reserved. No part of this publication may reproduced or transmitted in any form or by any means, electronic or mechanical, including photocopying, recording, or any information storage and retrieval system, without permission in writing from the publisher.

© Heinz Schumann, David Green and Chartwell-Bratt Ltd, 1994

German-English translation by Paul Green *B.A.*
Chartwell-Bratt (Publishing and Training) Ltd
ISBN 0-86238-373-0

Printed in Sweden
Studentlitteratur, Lund
ISBN 91-44-48661-8

Printing:	1 2 3 4 5 6 7 8 9 10	1998 97 96 95

Contents

PREFACE 5

CHAPTER 1
An introduction to Cabri-géomètre 9

CHAPTER 2
Learning geometry through interactive construction 31

CHAPTER 3
Creating macros for basic geometrical constructions 57

CHAPTER 4
Discovering theorems by varying geometrical figures 72

CHAPTER 5
Angle theorems as invariance properties 87

CHAPTER 6
Discovering theorems by generalising geometrical figures 108

CHAPTER 7
Producing and using loci 119

CHAPTER 8
Generating geometrical figures with line symmetry 142

CHAPTER 9
Solving advanced geometrical construction tasks 152

Contents

CHAPTER 10
Exploring drag-mode geometry   169

CHAPTER 11
Designing geometrical microworlds   204

CHAPTER 12
Microworlds for isoperimetric problems   232

CHAPTER 13
Microworlds for geometrical transformations   243

CHAPTER 14
Theorem finding – an advanced case study   250

CHAPTER 15
A geometric story   260

Appendix
Twenty problems to investigate   262

GLOSSARY   272

BIBLIOGRAPHY   275

Index   279

PREFACE

> *Man is a tool-using animal…*
> *Without tools he is nothing,*
> *with tools he is everything.*
> Thomas Carlyle

Geometry at a simple level is very important – length, area and angle are very basic concepts of wide applications. Ideas of symmetry are important in art, architecture, physics and chemistry. Geometry provides one of the best opportunities that exists to learn how to model the world and *do* mathematics – to distinguish among axiom, definition and theorem, to make conjectures and seek proofs or refutations.

However, over the past fifty years formal Euclidean geometry has almost disappeared from our schools. One reason for this was that in the past geometry has been very difficult for pupils to *explore* – they were simply given theorems to learn and later prove and apply to mostly meaningless problems. Robitaille and Travers (1986), reporting on a survey in North America, summarised the situation about ten years ago as follows:

> "The picture of the North American mathematics curriculum in geometry at the junior secondary or middle school level which emerges from an analysis of the data from the Second International Mathematics Study is not particularly encouraging. One is left with an impression of a lack of clear direction or progress toward an identified goal. There is geometry in the curriculum, but no one seems to be sure why it is there, how much time to devote to it, or how best teach it."

Now it *is* possible for students to explore Euclidean geometry *using computer software* so there is the chance for them to feel *ownership* of the subject matter – and, with guidance, to rediscover many important and elegant theorems for themselves. This may or may not lead them on to formal proofs – that is another (important) matter. The purpose of this book is to bring awareness of the tremendous possibilities of the new software to the attention of the interested teacher. There is the opportunity for Euclidean geometry to come alive again and, with the arrival of computer algebra to ease the burden elsewhere in the mathematics curriculum, there may become rather more time available at upper school level for doing *real* mathematics – and geometry has much to offer.

The content of this book is confined to plane geometry. But software development is going on apace and we will soon have 3-D systems (for example, a 3-D Cabri-géomètre) for teaching and learning spatial geometry at secondary school. The use of such tools will help us discover – and develop – Euclidean geometry in space which will, in comparison to plane geometry be much more revolutionary. Hardware development occurs at amazing speed and we dare not predict what geometric instruction will be like when students use note-book technology as their personal medium.

Recent recognition of these developments and their import for geometry education is to be found in a Mathematical Association pamphlet *Not the National Curriculum* (1992). In the chapter entitled *The Ideal Geometry Curriculum?* the late John Higgo wrote:

"The teaching of geometry has undergone a radical change in the last few decades. This change has, arguably, broadened the scope of geometry and made aspects of it more accessible to the majority. On the other hand the changes have created a loss of cohesion in the geometry curriculum, a loss of much traditional content and the virtual disappearance of the use of geometry as a vehicle for the acquisition of a logical approach to mathematics. In addition, recent development in technology require us to reconsider how geometry should be taught."

"It is because geometry is a useful model of the world around us that we are suggesting it as an essential part of our curriculum. The fact that it is a *model* is as important for mathematics as the fact that it is *useful*."

"The computer offers us a means of implementing our abstract model of the real work and it offers an aid to learning geometry."

Higgo proposed the following developmental objectives:

Pupils should develop

(a) independent thinking and logical thinking through solving problems and working on open-ended tasks.
(b) spatial understanding (including three dimensions)
(c) the ability to represent geometrical objects and to measure accurately using a variety of instruments, including traditional geometrical instruments as well as computer packages and such modern devices as electronic tapes, plumb lines, motion sensors and graph plotters
(d) knowledge and understanding of geometrical figures – both solid and planar
(e) knowledge and understanding of geometrical transformations and the ability to apply them
(f) an appropriate mathematical language and vocabulary
(g) awareness of links between geometry and the rest of mathematics, with other school subjects and with the real world
(h) the ability to think imaginatively
(i) the ability to formulate, test, generalise, extend and discuss conjectures
(j) a willingness to find and use their own methods to solve problems
(k) pleasure in shape and form and the mathematical ideas associated with them.

Higgo concludes by saying that

"The opportunity to DO, EXAMINE, PREDICT, TEST, GENERALISE should, from an early age, permeate the learning situations pupils are put in. They should be encouraged to question (WHY?) and extend (WHAT IF?) their findings. Geometry should be presented in such a way as to highlight the logical aspects. At appropriate stages, children should be helped to go on to formulate their own proofs (sometimes as a group). What is important, however, is that we do not restrict pupils' progress (denying them the opportunity to 'act as mathematicians') and do not try to separate or compartmentalise the stages too much."

The authors of this book heartily endorse all the above and present this book as a first step towards encouraging and supporting teachers to provide for their students an interactive computer geometry environment which can directly or indirectly achieve many of the above aims.

This book is a product of English-German cooperation – an initial step towards overcoming in a practical way the barriers of language and tradition which divide European national curricula.

Discovering Geometry with a Computer

Chapter 1 in particular, and Chapter 2 to some extent, provide an introduction to Cabri-géomètre. The remaining chapters are largely independent of each other and the reader may read them in any order.

There is a disc of Cabri-géomètre figures and macros for sale from the publishers to accompany this book. This disc, available in *Macintosh* and *MS-DOS* formats, contains over 400 files and macros enabling readers to work through the book without having to create the files and macros, and to examine the files (using the History command) to see the details of the constructions.

This book is not a course in geometry but rather an exploration of geometry using Cabri-géomètre. We hope you the reader will join us in this exploration, no doubt going your own way from time to time, as interesting avenues open up as they surely will ...

Acknowledgements

The authors wish to thank the following:

- Paul Green for providing translation services;
- William Wynne Willson for providing valuable advice on some of the material in this book;
- Nicci Davis, Joan De Souza, Louise Howard, Helen Sherwood, and Carole Torr for dealing so well and so cheerfully with a difficult manuscript;
- Philip Yorke of Chartwell-Bratt for encouragement and support;
- Monika and Valerie for their patience in the long hours when their partners were not around!

Heinz Schumann   David Green
Waldburg/Weingarten, Germany   Loughborough, England

August 1994

CHAPTER 1

An introduction to Cabri-géomètre

Euclid's revenge

1.1 INTRODUCTION

Geometry used to have pride of place in mathematical education but in the last fifty years its role has diminished and formal geometry has disappeared. There are three major reasons for this: firstly the difficulty of actually performing the necessary constructions accurately, secondly the considerable time consumed in repeating drawings, and thirdly the realisation that the proof concept fundamental to traditional Euclidean geometry is inherently difficult for most students and that parrot learning of proofs has no merit. However, new IT tools are now arriving, one of which is called Cabri-géomètre, which can address these issues and can make traditional geometry live again. The cry of the modern mathematics movement of the early Sixties was "Euclid must go!"; the cry of the Nineties could be "Come back Euclid!".

Cabri-géomètre has been likened to LOGO. This is an accurate comparison not so much because of the mathematical content as in the *approach* (for it must be doubted whether LOGO is an appropriate tool for exploring Euclidean geometry *per se*). Cabri-géomètre, like LOGO, has a "low threshold, high ceiling", being quite easy to begin to use but capable of considerable development and sophistication. It shares with LOGO the facility to build up more complex structures from simple basic 'objects' which then themselves become 'objects' of the enriched system. LOGO does this with *procedures*, Cabri-géomètre with *macro-constructions*. As with LOGO, it is perhaps unwise to ask (or at least to try to answer) the question "What can it do?". The response must be "It can do a great deal, and no one knows how much!".

1.2 WHAT IS CABRI-GÉOMÈTRE?

A team of professional programmers, computer scientists and mathematical educators, under the leadership of J-M Laborde at the University of Grenoble, have developed this graphics system for geometrical construction. Its international première took place at the 6th International Congress of Mathematical Education in Budapest in 1988. Its name is derived from un **Ca**hier de **br**ouillon **i**nteractif which means an interactive sketch book. It is written in the programming language C, using graphics utilities. The original implementation was written to blend in with the Apple Macintosh desktop-orientated user interface. It is also available for the IBM PC family. Cabri-Géomètre is created according to accepted software-ergonomic principles and is therefore quite easy to use.

The manual, to which the reader should refer for detailed technical information, says that Cabri-géomètre is "An interactive notebook for learning and teaching geometry". It takes the familiar *point*, *line* and *circle* as most basic **objects** and allows the user to draw geometrical shapes on the screen and to specify **relationships** between them (e.g. a point **on** a line, or a line **perpendicular to** another line). The drawn shapes can be moved and distorted and any defined relationships will be preserved. By this means the invariant features can be investigated, which is the essence of geometry. Hypotheses can be formed and tested visually *and* numerically *and* deductively and theorems 'proved' . Cabri-géomètre, which can be thought of as the geometrical equivalent to the spreadsheet, has potential from primary school through to university.

1.3 WHAT COMPUTER DOES CABRI-GÉOMÈTRE RUN ON?

Cabri-Géomètre was originally developed with a French language interface using the Apple Macintosh computer and is now available in English versions for the Macintosh family of computers and IBM PC compatible computers running MS-DOS. This introductory chapter describes version 2.1 for the Macintosh, which has some definite enhancements over version 2.0. A version for the Acorn Archimedes computer, popular in British schools, is being considered. In addition, Heinz Schumann, co-author of this book, has provided a German language interface for the Macintosh version.

An introduction to Cabri-géomètre

Cabri-Géomètre is distributed in the UK and many other countries by Chartwell Bratt (Old Orchard, Bickley Road, Bromley, Kent BR1 2NE, Tel. 081-467-1956) and by Brooks Cole in the USA.

There follows a short description of Cabri-géomètre, emphasizing some essential features and characteristics. The reader is also referred to the very useful manual provided with the system. The reader familiar with the system might prefer to skim or skip the rest of this introductory chapter.

1.4 GETTING STARTED (Creation)

When Cabri-Géomètre is loaded, the main menu appears across the top of the screen (Fig. 1.1). It comprises five pull-down menus. In this section we consider the **Creation** menu which is the first to use when drawing shapes.

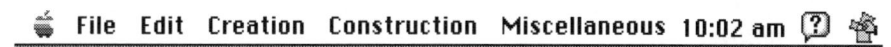

Fig. 1.1 Menu bar.

Cabri-géomètre takes as its fundamental objects

>   Basic point
>   Basic line
>   Basic circle

which may be drawn on the screen with simple commands from the **Creation** menu (Fig. 1.2). They appear as indicated in Fig. 1.3. (Note that different versions of the software use slightly different names for some menu options. For example **Point** and **Basic point** are both used.)

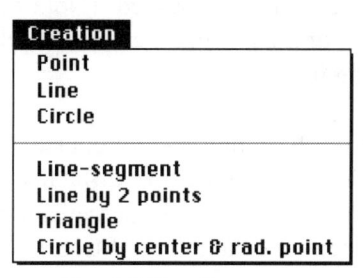

Fig. 1.2 Creation menu.

11

Discovering Geometry with a Computer

Point	Line	Circle

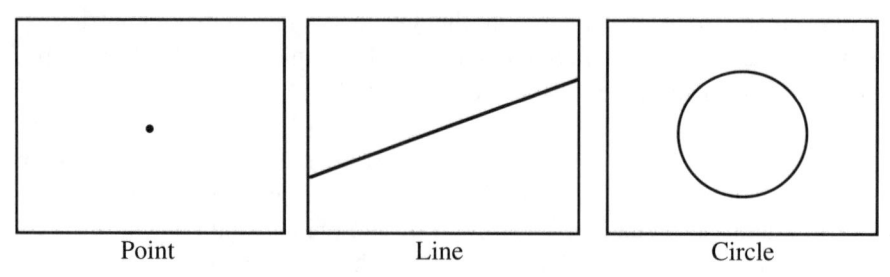

Fig. 1.3 The three basic Creation menu objects.

Of the four other **Creation** commands, three are fundamental:

> *Line by 2 points*
> *Circle by centre & radius point*
> *Line-segment*
> whereas the last
> *Triangle*

is available for convenience. Fig. 1.4 shows how these simple figures appear on screen.

Drawn objects can be moved about the screen (by 'dragging' with the mouse). The three basic shapes move as entities, but the four other shapes (defined in terms of points) are moved and deformed by the movement of each defining point. For example a circle defined using **Circle by centre & rad. point** may be displaced and enlarged by moving one or other of its defining points, and similarly a line segment may be lengthened and rotated. A circle defined simply by **Circle** can be moved about the screen but its size is fixed and there is no centre point or point on the circumference unless they are explicitly constructed. (Note that the size of a basic cicle and the orientation of a basic line *can* be altered by holding down the OPTION key while dragging.)

This much is not particularly impressive, of course, being the standard fare of drawing packages. Some packages have difficulty in ensuring that circles appear as circles (and not ellipses) on the screen and when printed out; fortunately this does not seem to be a failing of Cabri-géomètre.

An introduction to Cabri-géomètre

Line segment	Line by 2 points
Triangle	Circle by centre & rad. point

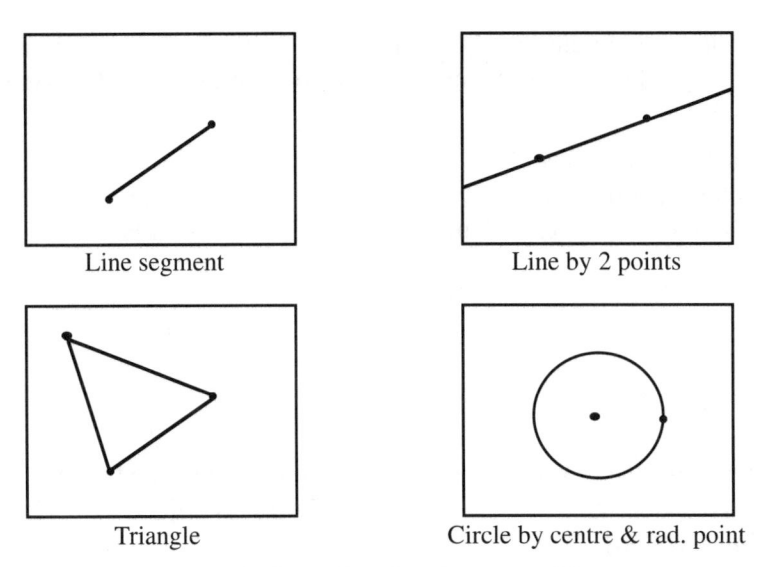

Fig. 1.4 The four other Creation menu objects.

1.5 DEFINING RELATIONSHIPS (Construction)

A very powerful feature of Cabri-géomètre is the ability to define *relationships between objects* and to explore graphically the implications. This is fundamental to Euclidean geometry and is likely to be the most used feature of this software at the school level.

The **Construction** menu has a number of commands which allow relationships between objects to be established (Fig. 1.5). Included is the facility to draw a locus of a moving point.

```
Construction
Locus of point

Point on object
Intersection

Midpoint
Perpendicular bisector
Parallel Line
Perpendicular line
Center of a circle

Symmetrical point
Bisector
```

Fig. 1.5 Construction menu.

The following simple example (Fig. 1.6) illustrates the kind of activity which is possible. It uses commands from both **Creation** and **Construction** menus.

1. Draw a circle defined by centre and radius point.
   [Circle by centre & rad. point]

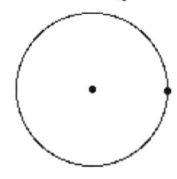

2. Draw a diametric line through the the two points
   [Line by 2 points]

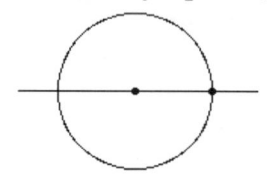

3. Determine the unknown endpoint of the diameter.
   [Intersection]

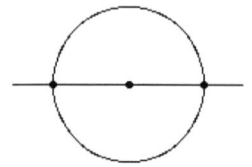

4. Define a free point on the circle.
   [Point on object]

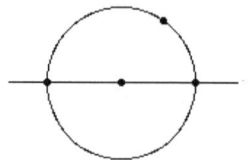

5. Draw a triangle whose base is the diameter and whose apex lies on the circumference.
   [Triangle]

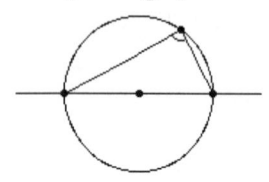

6. Look for anything interesting in the figure.

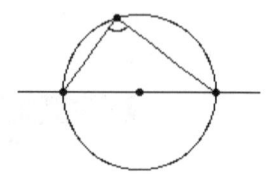

7. Form an hypothesis (e.g. the angle at the circumference is always right-angled).

8. Check the hypothesis visually by moving the free point around the circumference.

Fig. 1.6 A simple sequence of commands.

Such work could be carefully structured (contrived, some might say) or could be the outcome of very free exploration. As with LOGO, various modes and styles of working are possible, each with its own merits.

There are further more powerful ways to test an hypothesis which will be described later.

1.6 NAVIGATION

Users can always see where they are and which menu options are available (menu options currently unavailable for use are indicated by grey shading rather than appearing black). The bar positioned under the Main menu bar shows in its centre the name of the current drawing (Fig. 1.7). To the right of the name is a box to click for context sensitive help. To the left of the name is another box to click to stop an activity or to abandon, clear or delete the current figure.

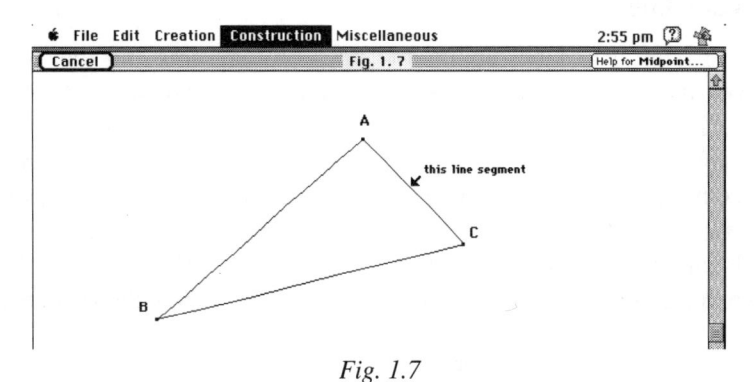

Fig. 1.7

1.7 DRAG-MODE GEOMETRY

Figs. 1.8-1.10 show how a drawing can be continuously transformed with the drag cursor. An acute-angled triangle ABC (Fig. 1.8) is transformed into a right-angled triangle (Fig. 1.9) by dragging vertex C, and then into an obtuse-angled triangle (Fig. 1.10). The two other basic points A and B remain fixed; the transformation is circle-invariant and straight-line-invariant and retains incidence, ordering and orthogonality relationships. Drag-mode enables the user to vary construction results 'continuously'.

Figs. 1.8–1.10

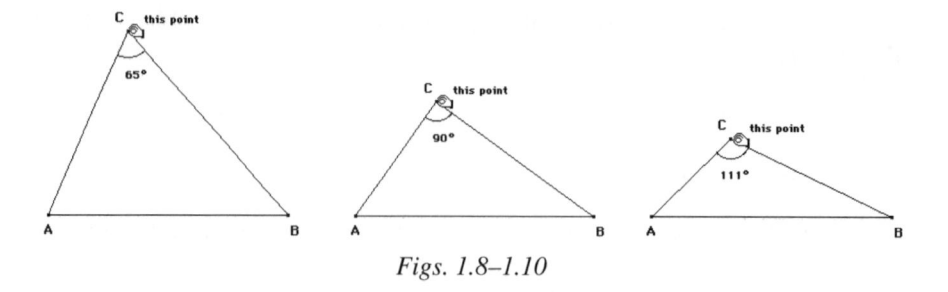

1.8 THE EDIT MENU

1.8.1 Labelling

It can be helpful to label points and lines on diagrams and this is easily achieved using the **Label** (or **Name**) command found in the **Edit** menu (Fig. 1.11). This was used in the previous example to label the vertices of the triangle (Fig. 1.8). Labelling is achieved by choosing the appropriate **Edit** command, pointing with the mouse cursor and clicking where the label is to go, and then typing in the desired text, which can be finely positioned with the mouse before fixing by pressing RETURN or clicking elsewhere. (Note: to re-position or amend the label repeat the sequence.)

Edit	
Undo	⌘Z
Cut	⌘X
Copy	⌘C
Paste	⌘V
Clear all	
Look of objects	⌘G
Enlarge the figure	
Name	⌘L
Comments...	⌘B
Show page...	
Preferences...	

Fig. 1.11 Edit menu.

1.8.2 Altering the look of objects

A very useful feature when displaying figures is the **Look of objects** command found in the **Edit** menu (this is called **Appearance of objects** in some versions of Cabri-géomètre). With this one can either hide or emphasize lines. As an example, we take the construction of the centroid (G) and orthocentre (H) of a triangle. To construct these important points requires a number of intermediate (or auxiliary) point and line constructions (Fig. 1.12).

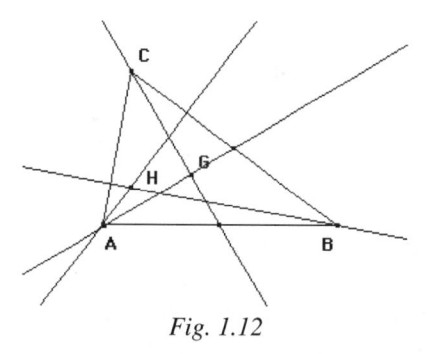

Fig. 1.12

These intermediate objects – the midpoints of the sides and the corresponding medians, and the perpendiculars to the sides passing through the vertices – can be hidden in the final diagram. This is accomplished by using **Look of objects** and selecting the erase pen icon and clicking on the objects. While **Look of objects** is active they appear dotted (Fig. 1.13).

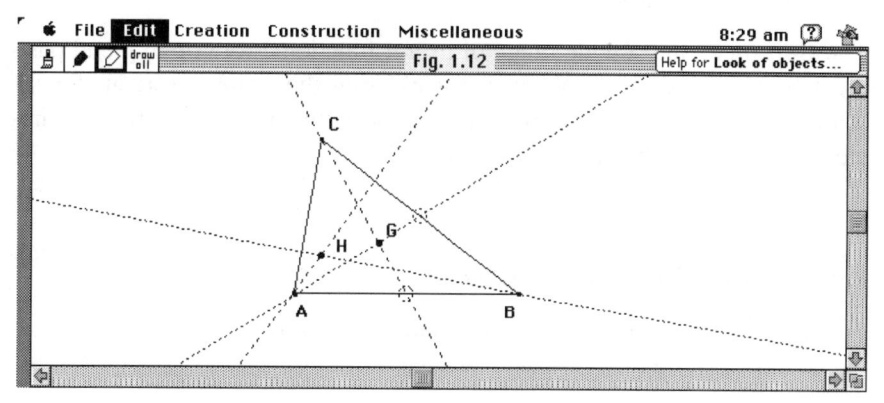

Fig. 1.13

As well as hiding objects we can emphasise objects. We designate as most important the orthocentre (H) and centroid (G). These points we cause to appear in bold by choosing the paint brush icon and clicking on them. When this process is completed and **Look of objects** is quit the unwanted lines disappear, the two points appear larger and the drawing is considerably clearer (Fig. 1.14).

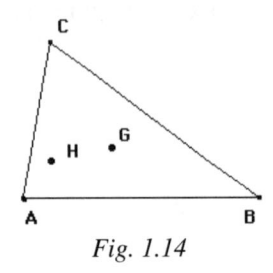

Fig. 1.14

The figure can be varied by dragging any of the vertices and the effect on the centroid (G) and orthocentre (H) observed. (For which triangle do the two points coincide? Can H be made to coincide with a vertex? What effect does doing this have on the triangle's shape? Can this be done for G?).

With complex figures in particular, it would be very helpful to be able to use different colours but that is a facility not available in current versions of Cabri-géomètre, unfortunately.

1.8.3 Other commands

The **Edit** menu has a wide range of useful functions: for copying and pasting figures (using the clipboard); for clearing the graphic screen; for rescaling figures; for naming objects; for the adjustment of system parameters; for undoing the previous command.

1.9 THE MISCELLANEOUS MENU

This menu too has many useful commands (Fig. 1.15).

An introduction to Cabri-géomètre

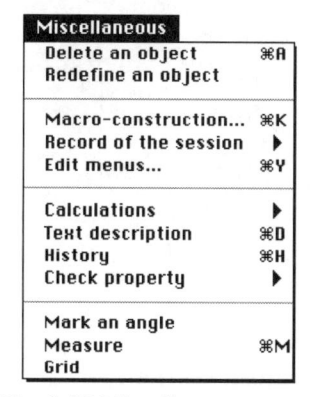

```
Miscellaneous
Delete an object          ⌘A
Redefine an object

Macro-construction...     ⌘K
Record of the session     ▶
Edit menus...             ⌘Y

Calculations              ▶
Text description          ⌘D
History                   ⌘H
Check property            ▶

Mark an angle
Measure                   ⌘M
Grid
```

Fig. 1.15 Miscellaneous menu.

1.9.1 Marking an Angle

It is very useful to be able to place an arc on an angle, as was done for example in Fig. 1.8. This is achieved by using the **Miscellaneous** menu option **Mark an angle**.

1.9.2 Measuring

Although measurement of lengths and angles in standard units is not fundamental to this approach to geometry there is a clear use for it, and the **Miscellaneous** menu option **Measure** allows this.

The measurement of line length and angles is an excellent tool available in Cabri-Géomètre because it is dynamic i.e. as the figure moves the measurements are automatically updated on screen. An earlier example (Figs. 1.8-1.10) illustrates this. A further example is given here – the experimental construction of a regular pentagon.

Any pentagon is drawn by joining line segments to five points. The pentagon's internal angles are marked and measured (Fig. 1.16). By dragging the vertices in turn the sides are made all the same size (Fig. 1.17) and further dragging makes the angles equal too, producing a 'regular' pentagon (Fig. 1.18). Almost any geometrical construction task requiring lengths or angles with given values can be experimentally solved in this way.

Figs. 1.16–1.18

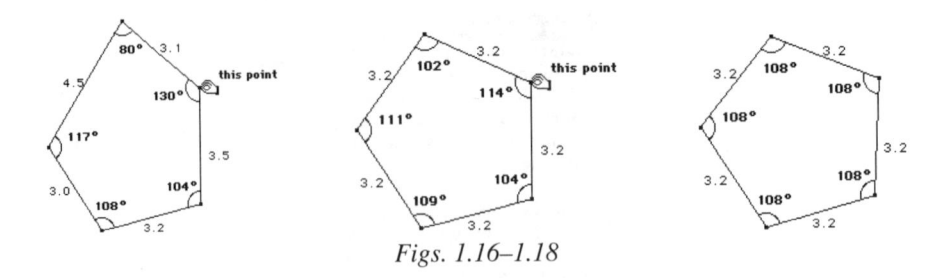

1.9.3 Calculations (Macintosh version)

Further calculation facilities are available from the **Miscellaneous** menu and appear in a temporary window. These include: Coordinates of a point; Area enclosed by a polygon; and Slope of a line. Unfortunately the **Calculations** window usually disappears as soon as adjustments are made to the drawing (it is moved behind the graphics window) but it can be recovered by juggling with the positions of the windows on the screen or by selecting the **Show calculations** option from the **Calculations** submenu of the **Miscellaneous** menu. A good use of this is to demonstrate Pythagoras' Theorem (Fig.1.19). When the triangle is deformed the calculations are automatically updated in real time.

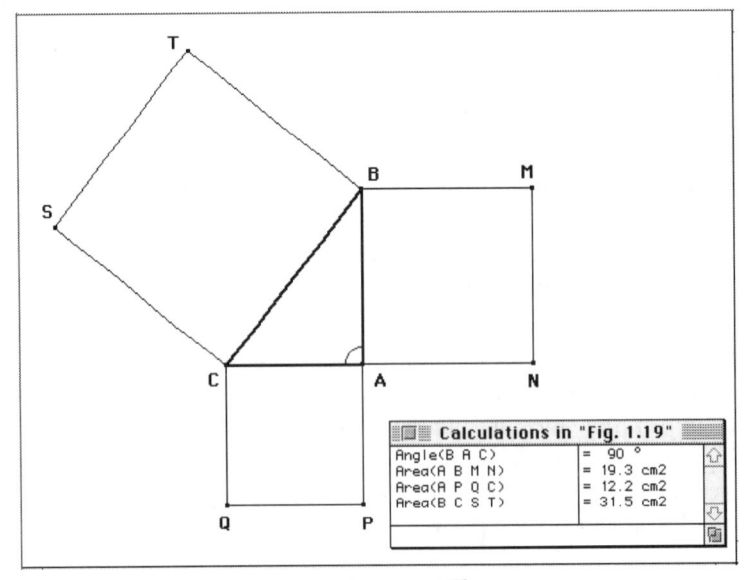

Fig. 1.19 Pythagoras' Theorem.

1.9.4 Other commands

The **Miscellaneous** menu has many other useful functions: for deleting individual objects; for altering relationships between objects; for defining new constructions; for customising the system by specifying the allowable basic objects and constructions (mainly so that the teacher can adapt the system to the curriculum and competence of the students); for the replaying of a previous working session (**Record**); for the repetition of a particular construction sequence (**History**); for placing an arc on an angle (**Mark an angle**); for overlaying the screen with 1cm squares (**Grid**).

1.10 TOWARDS PROOF (Macintosh version)

The ability to calculate and display measures of angles, lengths and areas can corroborate numerically a visual impression (such as an angle being constant at 90°, or one length always being twice another). For the triangle-in-the-circle example discussed earlier (Fig. 1.6) the next step might be:

Action
9. Check the hypothesis numerically by measuring the angle and moving the free point. Also, check further by altering the circle itself by moving the points defining the centre and radius.
   (See Fig.1.20.)

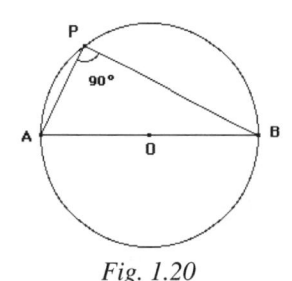

Fig. 1.20

There is a further and most significant step towards a complete confirmation of the validity of an hypothesis, and confirmation that it is a theorem. This uses the **Check property** command found in the **Miscellaneous** menu.

Action
10. Prove or refute the hypothesis by asking Cabri-géomètre if it is generally true.

[Check property–*perpendicular lines*]
Click on the two lines which are thought to be perpendicular. (The software responds with a message – see Fig. 1.21).

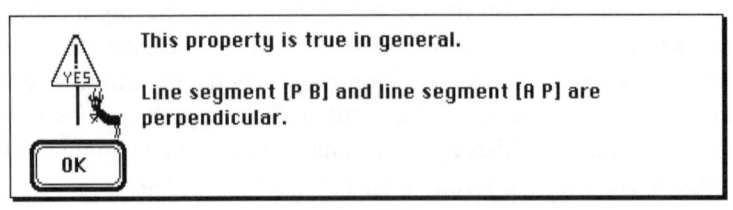

> YES
> This property is true in general.
>
> Line segment [P B] and line segment [A P] are perpendicular.
>
> OK

Fig. 1.21

How does it know? Not by having axioms and theorems stored but rather by means of a semi-intelligent system which varies the figure's variables and checks numerically. If the property *is not* generally true then that is reported. If the property *is* true in the particular diagram but not generally true then the diagram is modified (e.g. a point or line is moved) to provide a counter-example. For example, a general triangle ABC is drawn such that it *appears* isosceles (Fig. 1.22).

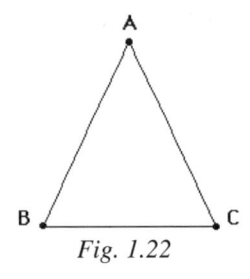

Fig. 1.22

Then an enquiry is made as to whether the two sides AB, AC, are equal (which they are in the particular figure). The response is shown in Fig. 1.23.

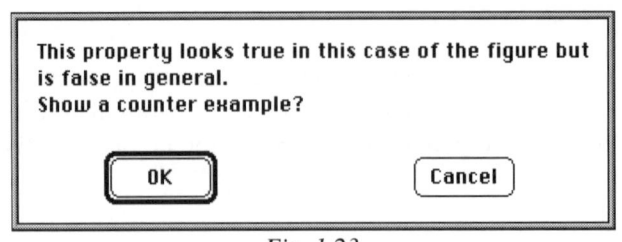

> This property looks true in this case of the figure but is false in general.
> Show a counter example?
>
> OK   Cancel

Fig. 1.23

22

If a counter-example is requested Fig. 1.24 is obtained. The position of B is moved and the old position of AB is shown as a dotted line and the two lengths AB, AC (no longer equal) flash.

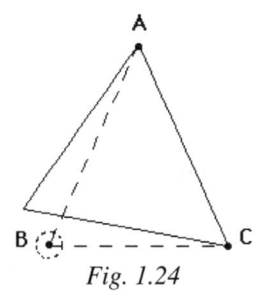

Fig. 1.24

In principle, *any* theorem of two dimensional Euclidean geometry can be investigated, illustrated, tested and refuted (by counter-example) or (arguably) confirmed. This of course raises difficult questions about *what is proof* and about how it should be treated in the mathematics classroom. It may be helpful to think of Cabri-géomètre as confirming whether or not it is worth trying to prove a conjecture by more rigorous means, rather than considering that the software actually provides the proof.

1.11 LOCI

The locus command enables paths of moving points to be traced on the screen (and there are printing facilities too, of course).

An example is given here to find the path of a point equidistant from a fixed point and from a point constrained to move around a fixed circle. The sequence of steps is given below. All have corresponding commands in the Cabri-géomètre menus; points and lines are positioned by simple mouse actions.

1. Draw the fixed (basic) point, P.
2. Draw the fixed (basic) circle, c.
3. Place the free point, Q, on the circle c.
4. Find the midpoint, M, between the two points P and Q (Fig. 1.25).

Discovering Geometry with a Computer

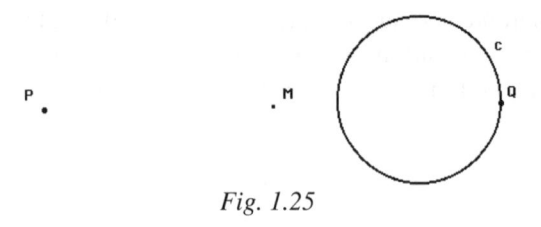

Fig. 1.25

5 Ask for the locus of M.
6 Move the free point Q round the circle c and so obtain the locus of M.
7 Print the resulting diagram (Fig 1.26).

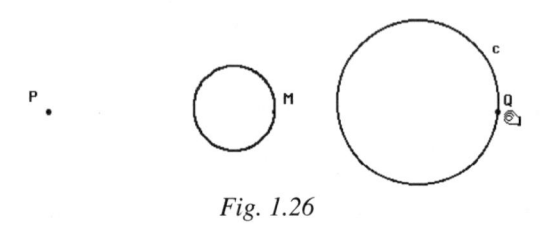

Fig. 1.26

The reader might like to consider how to obtain the locus of a point moving equidistant from a fixed point P and a fixed straight line m. (Fig. 1.27 shows the result where Q is a free point moving on the fixed line m.)

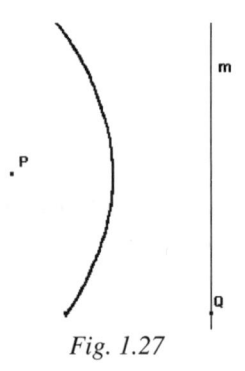

Fig. 1.27

Fig. 1.28 records the dynamic generation of a geometric locus producing a curve called the Trisectrix. This is the path followed by the orthocentre (H) of a triangle (ABC) when a vertex (C) is dragged in a circular path around another vertex (A). Using Cabri-géomètre, the relationship of the circle to the triangle must first be established and the vertex C fixed onto the circle; there are several cases to be distinguished. This example is dealt with more fully in Chapter 7, Section 7.4.

An introduction to Cabri-géomètre

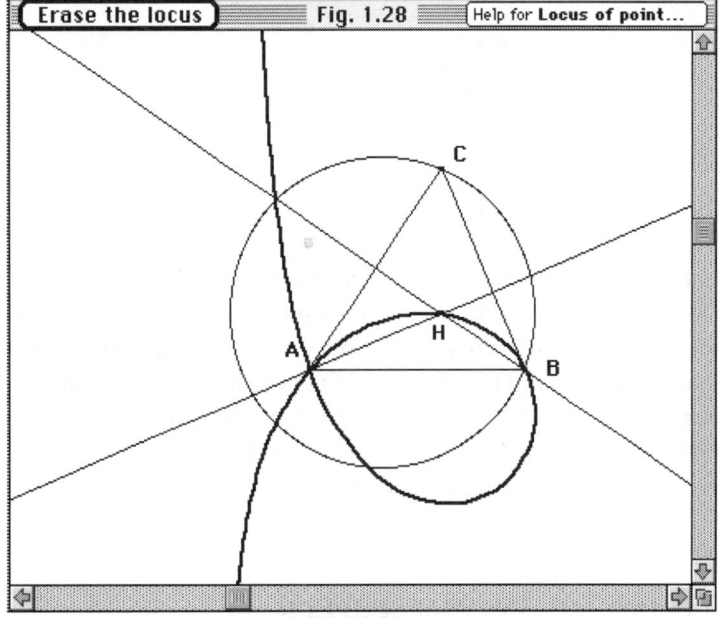

Fig. 1.28

A limitation of the software – hardly a surprising one – is that the drawn locus is no more than a trace on the screen. The trace itself cannot be manipulated or tested in any way. Nevertheless the idea of locus is a powerful one and Cabri-géomètre complements other software already available for such explorations. Future versions of Cabri will resolve this difficulty.

1.12 DEPENDENCE AND INDEPENDENCE

Ideas of dependence and independence can be explored by establishing relationships among points on figures. A simple example is for one person unobserved by the rest of the group to place three points on the screen, A, B, C. The midpoint of B and C is found and labelled D (see Fig. 1.29). Then the question is asked "What is the relationship between the points?", and this can be deduced by moving the points. Using Cabri-géomètre the points A, B and C can all be moved independently but D will move if and only if B or C is moved. D is *dependent* on B and C and *independent* of A.

25

.A   .C
        .D

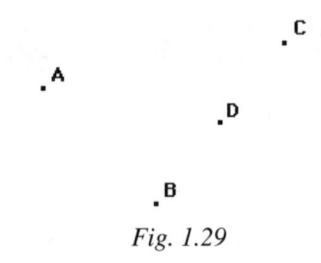

   .B

Fig. 1.29

The nature of the dependence is simple in this case but can be made more subtle, of course. For example, in the example above the point B could be hidden (using **Look of objects** or **Appearance of objects** from the **Edit** menu); alternatively, the midpoint of AD could be found and labelled E and D hidden. Figures involving points of intersection of pairs of lines or of lines with circles (possibly with the lines and circles themselves hidden) can provide interesting and demanding challenges.

1.13 MAKING THE SYSTEM SIMPLER

The supplied system has seven **Creation** menu commands (described already) and ten **Construction** menu commands. Of those ten, we have already met:

> **Locus of point**
> **Point on object**
> **Intersection (of two objects)**
> **Midpoint**
> **Centre of a circle**

The others available are:

> **Perpendicular bisector**
> **Parallel line**
> **Perpendicular line**
> **Symmetrical point**
> **Bisector (of angle)**

This may all seem too much! Fortunately, there is a facility to remove commands from menus to create a simpler system, and so tailor it to the user's needs. For example the two main menus might be restricted to:

Creation menu	Construction menu
Point	**Locus of point**
Line-segment	**Point on object**
	Intersection
	Midpoint

This would limit investigation to polygonal shapes, without parallelism or perpendicularity being available for construction. This menu restriction is achieved using the **Miscellaneous** menu option **Edit menus**.

1.14 BUILDING UP THE SYSTEM

The **Creation** and **Construction** menus provide the basic building blocks for geometrical investigation. As has been mentioned, **Triangle** is provided as a **Creation** command. This is not strictly necessary as a triangle can easily be constructed from three line segments. Triangle is, then, just a convenient extra. How about having an equilateral triangle or a square or a general rectangle or ....? Indeed it could be very convenient to have many other shapes available, particularly for specific topic work. This can be achieved by means of *macro-constructions*. A macro-construction is essentially a prepared procedure written once (e.g. by the teacher) and added, perhaps permanently, to the **Construction** menu. Creating a macro-construction is not a trivial exercise but it can be accomplished with a little forethought, care and practice.

As an example of the definition of a macro-construction (called macro for short), we construct the orthocentre of a triangle (the orthocentre is the point where the altitudes of the triangle meet; an altitude is the line from the vertex of the triangle which meets the opposite side at right-angles). When actually used, the macro must be supplied with three points representing a triangle's vertices, and it responds by supplying the orthocentre.2727

- First, three points are generated and a triangle formed, then perpendiculars are dropped from two vertices to the opposite sides and their point of intersection defined (Fig. 1.30).
- Then the **Macro** (or **Macro-construction**) command is chosen and the initial (given) objects (i.e the three points defining the triangle) are indicated using the mouse (Fig. 1.31).

Fig. 1.30 *Fig. 1.31*

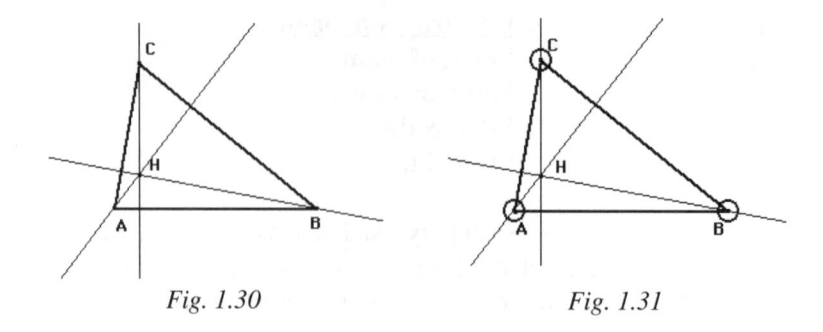

- Then the final (target) object – the orthocentre, H – is indicated (Fig. 1.32).
- Finally, the construction is given a name (Fig. 1.33).

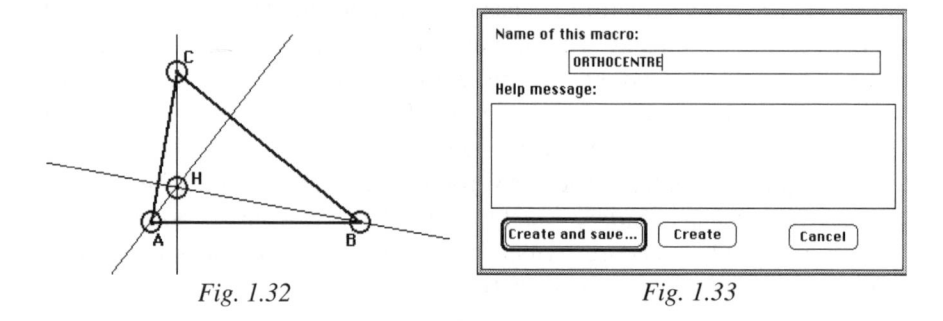

Fig. 1.32 *Fig. 1.33*

- When this process has been completed, the name of the newly defined construction appears at the end of the **Construction** menu (Fig. 1.34).

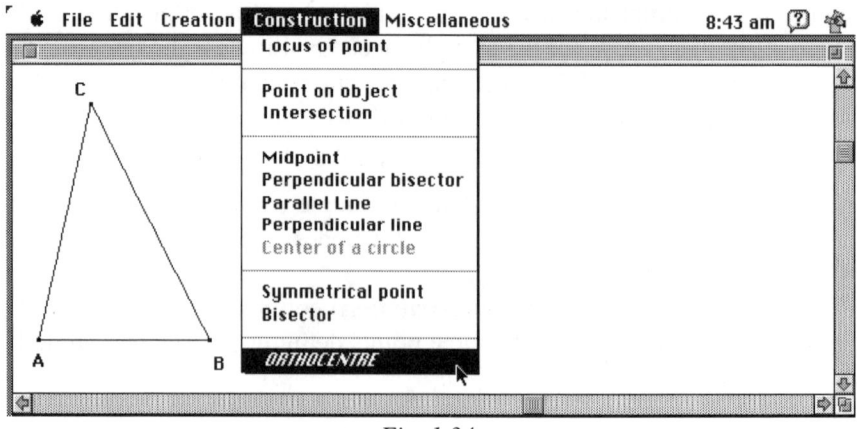

Fig. 1.34

This macro can now be applied to *any* 3 points and will produce the ortho-centre (in the case of 3 collinear points, a suitable warning message appears).

Thus, Cabri-géomètre is 'teachable' and extendable; it can 'learn' new constructions in an interactive way.

We see, then, that the menus can be customised to suit the situation and this can greatly help beginners to make progress without the distractions of many commands on the menus or without the need to construct a shape fundamental to a particular exploration. Another attractive feature is the ability to work with different subsets of commands – equivalent to having different restrictions on drawing apparatus.

1.15 FURTHER FEATURES ...

The briefest of mentions is given here of two other features which underline the similarity in philosophy to LOGO – they are the **Exposition** command which displays in a window, in natural language, a list of the objects shown on the screen, and the **History** command which allows the user to reconstruct a completed figure step by step by successive clicks of the mouse button.

The Cabri-géomètre microworld allows the exploration of any aspects of mathematics capable of a geometric interpretation. This not only includes such obvious ideas as the properties of circles and triangles and their inter-relationships and associated loci, but also symmetry, transformations, tessellations, projections, location problems, trigonometric relationships, tangents and gradients, and even some aspects of arithmetic and the number line and algebra. (However, it should be noted that this geometrical microworld only allows 'ruler-and-compasses' constructions, which is a definite restriction, plus drag-mode – allowing deformation of figures which brings dynamism.)

Unfortunately, the version currently available does not conform totally to school geometry: it cannot create semi-infinite straight lines nor arcs of circles, and although the creation and measurement of non-orientated angles is possible, reflex angles cannot be measured. Direct numerical input of line length and angle size is not supported. However, the development of a calculation module, compatible with the graphics of the system, has recently been developed and will appear in a new version.

Cabri-géomètre encourages emphasis on the process of doing mathematics, and on the exploration of the nature of mathematical proof. All-in-all Cabri-géomètre and similar packages can give us the best of both mathematical worlds – both 'traditional' and 'modern' . If past experience is anything to go by this software will be used in ways as yet unimagined!

We have only considered more elementary configurations here. Fig. 1.35 shows a typical advanced configuration typical of work found in later chapters. In the remainder of this book we attempt to convey an idea of the richness of the Cabri-géomètre microworld. Its potential remains a matter for speculation ... and exploration ...!

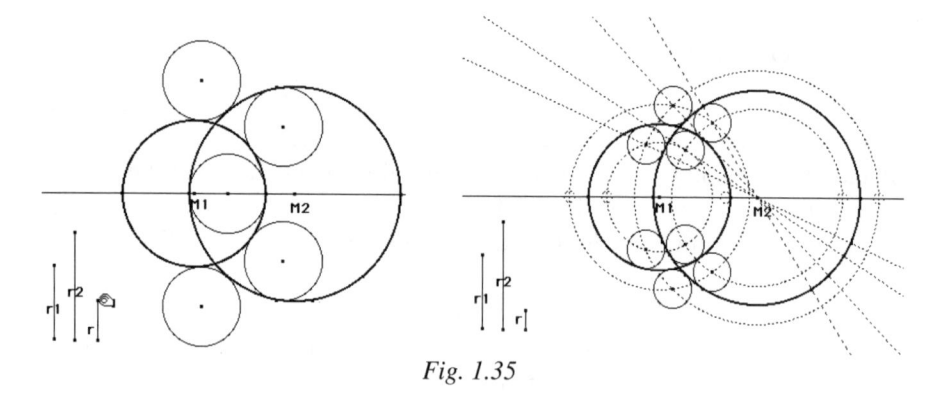

Fig. 1.35

CHAPTER 2

Learning geometry through interactive construction

We are restricted by the tools we use.

2.1 INTRODUCTION

This chapter presents some theoretical considerations and also provides a number of illustrative examples. As is the case throughout the book, the reader may find it beneficial to return to the more abstract sections having gained further practical experience through studying the many examples, preferably using the software itself.

An important way of learning geometry is through active exploration and reconstruction of geometrical systems. A system which provides the means of geometrical exploration and reconstruction, embracing abstract or practical techniques, is a valuable component of an active learning environment involving "learning by doing" (see Diagram 2.1). Such a system must be carefully developed by mathematical educators, and access to it needs to be more or less under the direct guidance of the teacher. What can be accomplished depends upon the tools and the media.

```
                    Means of exploration          ╱ System of: ╲
   ┌─────────┐    ─────────────────────────→    │   concepts   │
   │ Learner │                                  │  statements  │
   └─────────┘    Means of  reconstruction      │   theorems   │
                                                 ╲  operations ╱
```

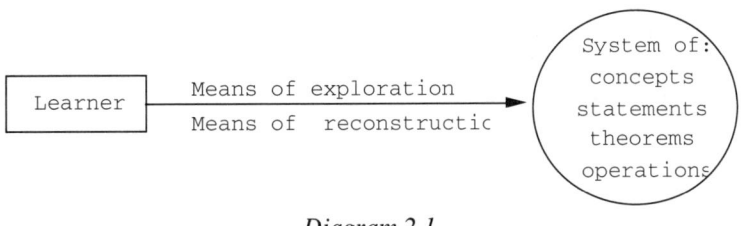

Diagram 2.1.

31

One means of exploration and reconstruction, which to a large extent determines the usual learning of elementary geometry in schools, is represented by the traditional analogue tools: compasses, straight-edge, set-square, ruler, protractor, paper and pencil. These conventional graphical construction tools have significant deficiencies as a means for the investigation and reconstruction of synthetic elementary geometry in schools. They provide only *limited* support for the economical acquisition of knowledge, for training in flexible and functional thinking, and for the development and application of intellectual techniques and heuristic strategies.

These deficiencies might be made good with the use of suitable interactive graphics systems. (We call such a system 'interactive' if a geometric construction can be worked through step by step controlled by command inputs from the user with corresponding step by step command execution by the system.) Current hardware and software limitations and economic constraints in schools mean that for now we must limit ourselves to 2-D elementary geometry and leave 3-D work to the future. The use of such interactive graphics systems can only supplement and not replace the conventional tools, for the following reasons:

- in the last analysis we do not know what importance derives from tactile contact with traditional analogue tools in the learning of basic geometry (it is impossible to conduct an experiment in which the effect of geometry learning by computer *alone* is investigated – students always come with prior knowledge and experience, none is a *tabula rasa*);
- working with the conventional construction and measuring tools represents a cultural technology, which is significant beyond the mathematical context;
- a worldwide, non-verbal standard of communication is inherent in the simple analogue tools;
- the continuity of the school curriculum must be kept in mind;
- hardware and software costs cause considerable problems (e.g. 'homework' presupposes students to have corresponding equipment available at home);
- the analogue tools are indispensable for practical applications (many people engage in DIY which must involve some practical geometrical activity from time to time);
- the definition of construction in geometry is to a large degree founded on ruler-and-compasses construction.

Geometric learning in schools today involves primarily:

- the measuring of lines, angles and polygonal areas,
- inductive learning of theorems and the formation of concepts,
- the solving of plane geometry construction tasks,
- and possibly, the construction of loci.

'Theorem-demonstrating examples' using paper or blackboard are, of necessity, *static* diagrams whereas a suitable interactive 2-D graphics system can be *dynamic* and support active exploration of elementary plane geometry.

2.2 DRAG-MODE

Drag-mode is an essential prerequisite for almost all interactive construction activities which go beyond the simulation of pure compasses, ruler and set-square constructions. With the drag cursor (shown in Fig. 2.1 as a grasping hand) we can freely move the basic objects of a construction, i.e. the points, lines or circles which determine the construction, and thereby transform the figure. Such transformations – apart from certain degenerative cases – preserve straight lines and circles.

Example 1

Figs. 2.1–2.3 illustrate the variation of a triangle drawn with two medians, showing that the medians trisect each other.

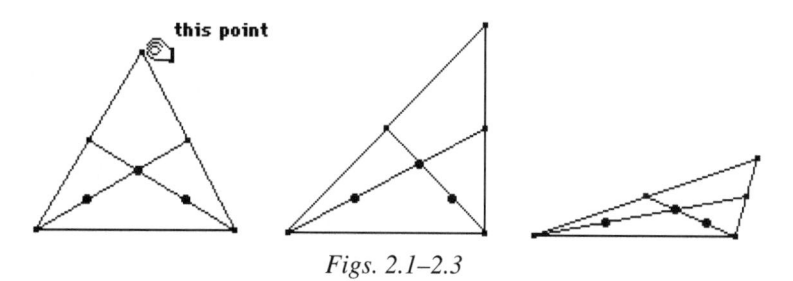

Figs. 2.1–2.3

The following relationships are generally invariant under drag-mode transformations:
- parallelism
- orthogonality
- proportionality of lengths (ratio)

- point symmetry (rotation through 180°)
- straight line (reflective) symmetry.

(This is provided that the relationships have been constructively established and are not mere accidents of the particular figure on screen.)

Drag-mode geometry can therefore be described as a hybrid of affine geometry and similarity geometry.

Incidence relationships may not be preserved, as illustrated in the following example.

Example 2

A line m is drawn passing over a circle c. The two points of intersection are obtained using **Intersection** from the **Miscellaneous** menu and labelled P and Q (Fig. 2.4). As the line is dragged with the mouse cursor, the two points become one (Figs. 2.5–2.6) and then disappear (Fig. 2.7) and the incidence property is temporarily lost.

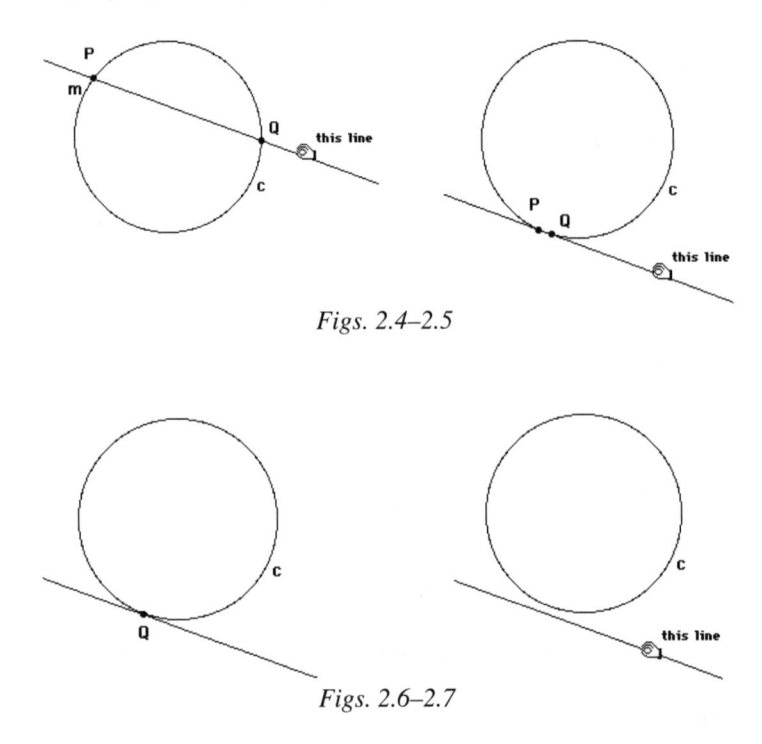

Figs. 2.4–2.5

Figs. 2.6–2.7

Drag-mode applied to a polygon can have different effects depending on the constraints of the basic objects which determine the polygon. The effect can be:
- a same orientated congruent transformation (shape, size and sense preserved)
  (e.g. translation or rotation);
- a same orientated similarity transformation (shape and sense preserved)
  (e.g. enlargement with positive scale factor);
- a (local) axis affine transformation (local, because only the particular figure being dragged is affected and not the whole plane)
  (e.g. equilateral triangle distorted to a scalene triangle).

A polygon defined in terms of linked line segments can be freely varied by dragging any vertex.

2.3 INDUCTIVE LEARNING OF THEOREMS AND CONCEPTS

Weaknesses of conventional construction methods for inductive theorem learning are:

- it is very time-consuming to produce a large enough number of suitable accurate configurations to represent the relevant theorem;
- it is only feasible for theorems based on simple configurations;
- it is not easy to avoid inaccuracy in measurements;
- only static configurations are possible, except perhaps in the simplest of cases (movement and variation being accessible only through mental imagery.)

Underlying much traditional geometry pedagogy is the *Pictorialistic hypothesis*:

> Mental images of figure transformations are formed analogously to the real transformations of physical models of figures.

Research has validated this but nevertheless the problem has remained that teachers have had to rely on students being able to visualise transformations mentally. This assumption has been necessary for both basic shape concepts (e.g. "What is a parallelogram?") and basic theorems (e.g. "A diameter of a circle subtends a right-angle at the circumference.").

Discovering Geometry with a Computer

A new situation now arises with Cabri-géomètre and similar systems which have drag-mode which have the important property that the transitions occur *continuously* in *real time* through *user-controlled* cursor movements. Thus the process of *acquiring knowledge* about a configuration can, to a large extent, be decoupled from the process of its *actual construction*, by providing the user with prepared configurations which can be easily varied. (The student loads the figure and manipulates it without having to construct it in the first place.)

Principle of configurative mobility
This principle states that inductive theorem finding in geometry is made easier by providing an interactive system which supports continuous variation of configurations. A system to enable this principle to be put into effect should allow the user:

– to produce many isomorphic configurations from one configuration (in real time);
– to produce special cases of a configuration out of a general case;
– to produce special cases of a configuration out of other special cases;
– to produce limiting cases of a configuration.

The operative question is: "Which characteristics of a configuration remain invariant, and which change, during continuous transformation?" Basic geometric theorems arise as statements of invariance when studying such transformations. The following selected examples illustrate the above ideas.

Example 3 *Position of circumcentre of a triangle*
In Figs. 2.8–2.10 the continuous variation of a triangle around a fixed circumcircle is shown. The circumcentre M is the intersection point of the three perpendicular bisectors of the sides of the triangle. M either lies inside the triangle (Fig. 2.8), on one of its sides (Fig. 2.9), or outside the triangle

(Fig. 2.10), according to whether the triangle is acute-angled, right-angled or obtuse-angled.

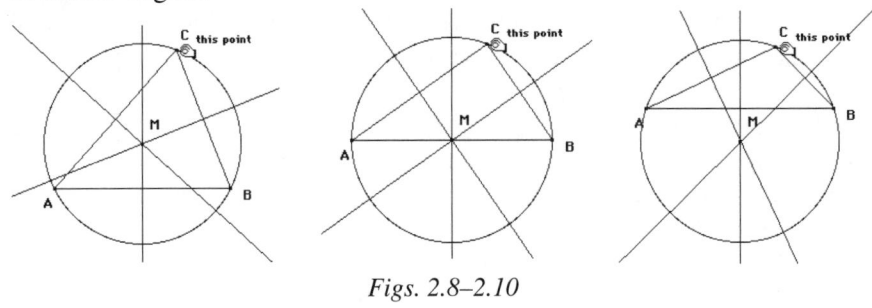

Figs. 2.8–2.10

Example 4 *Sum of distances from a point to the vertices of an equilateral triangle*

In Figs. 2.11–2.16 we can see that the sum of the distances from any point inside an equilateral triangle to the three sides is a constant equal to its height (6.0 in the example). Figs. 2.11–2.12 shows two general cases, and Figs. 2.13–2.16 show special cases, verifying this result which is known as Viviani's theorem.

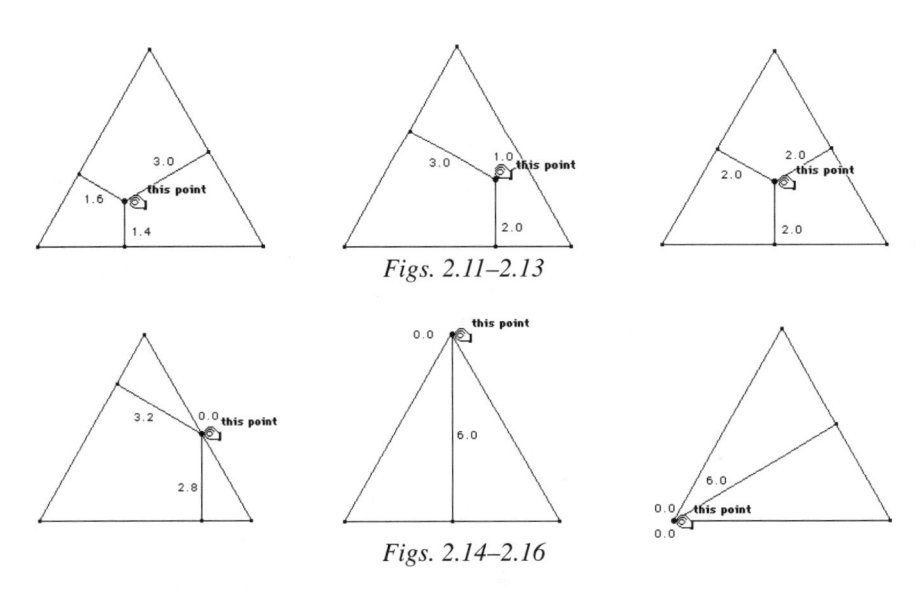

Figs. 2.11–2.13

Figs. 2.14–2.16

(What happens if the point lies *outside* the triangle? The interested reader can explore this – it is dealt with in Chapter 4.)

Discovering Geometry with a Computer

Example 5 *Quadrilateral formed by centres of squares on the sides of a quadrilateral*

Squares facing outwards are constructed on the four sides of a general quadrilateral (Fig. 2.17). This can be achieved by standard techniques or by using a macro, perhaps prepared by the reader or taken from the disc associated with this book (macro SQUARE). The centres of the four squares are joined to form another quadrilateral (Fig. 2.18). We vary the base quadrilateral and observe the effect on the centre-quadrilateral (formed by joining the centres of the four squares). We notice that the two diagonals of the centre-quadrilateral seem always to be equal in length and to intersect at right-angles (Fig. 2.19). This we confirm using the **Miscellaneous** menu option **Check property** if available.

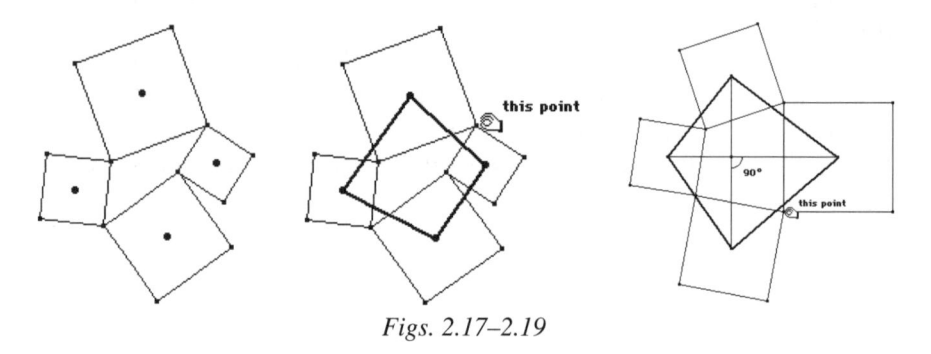

Figs. 2.17–2.19

As the base quadrilateral tends to the shape of a parallelogram, the centre-quadrilateral appears to take on the shape of a square (Fig. 2.20). We check the angles and sides and adjust the figure until the measurements show the centre-quadrilateral to be an exact square (Fig. 2.21). Then the angles of the base quadrilateral are checked to confirm that it is indeed a parallelogram (Fig. 2.22), establishing Thébault's theorem.

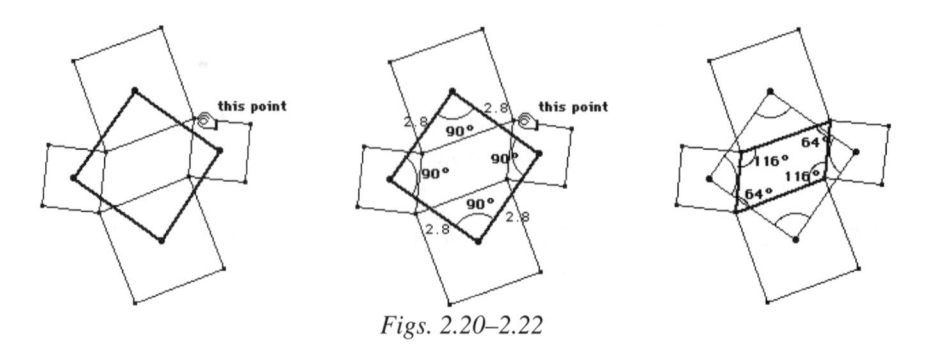

Figs. 2.20–2.22

Having worked through this general case we specialise and undertake a fresh investigation.

- We begin by constructing as base a parallelogram rather than a general quadrilateral.
- Then the centre-quadrilateral is constructed, as before (Fig. 2.23).
- Then we vary the base parallelogram.

We discover that the centre-quadrilateral remains a square, even for a degenerate parallelogram (Fig. 2.24). Turning the parallelogram 'inside out' has the result of turning the four side squares inwards (now overlaying the initial parallelogram); the centre-quadrilateral still remains a square (Fig. 2.25). What we have discovered here is an instance of the Napoleon-Barlotti theorem (for $n=4$). This states that the n-sided polygon formed from the centres of regular n-sided polygons constructed on the sides of an affine-regular n-sided polygon, is regular.

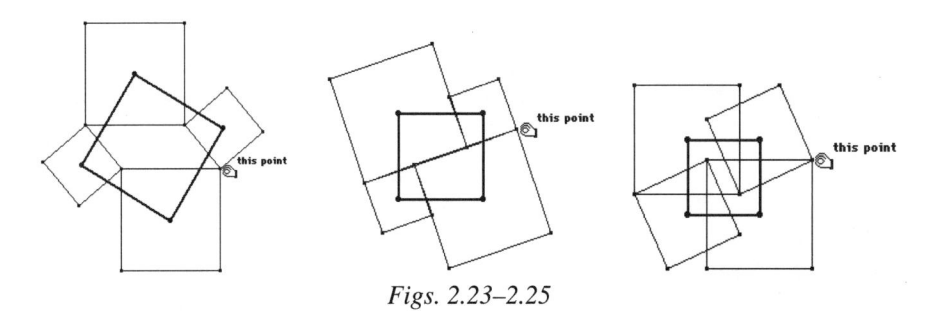

Figs. 2.23–2.25

2.4 MEASURING LINES, ANGLES AND AREAS

Weaknesses in conventional school geometry in measuring of lines, angles and (polygonal) areas are:

- all measurements are approximate and prone to error – especially so for angle measurement;
- calculating polygonal areas can be very time-consuming and so is rarely attempted;
- it is almost impossible to address questions aimed at finding functional connections, because the figures to be measured are static;

Discovering Geometry with a Computer

However, Cabri-géomètre enables the automatic measurement of lines, angles and polygonal areas and updating during continuous variation of objects in drag-mode. This opens up new possibilities, some examples of which are now presented.

Example 6 *Regularising a quadrilateral*
A general quadrilateral (Fig. 2.26) is changed, by dragging vertices, to produce in turn a parallelogram (Fig. 2.27), then a rhombus (Fig. 2.28), and finally a square (Fig. 2.29).

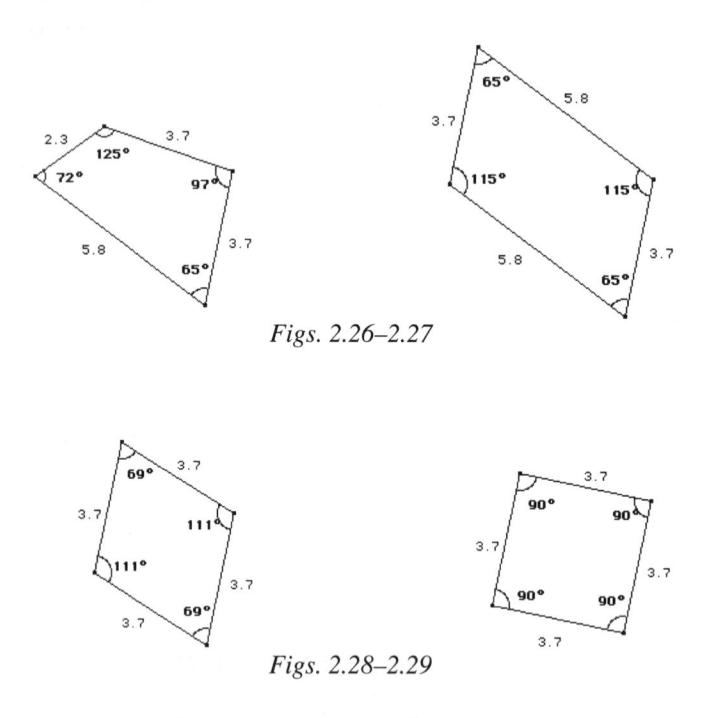

Figs. 2.26–2.27

Figs. 2.28–2.29

Example 7 *Angle-Side-Angle triangle problem*
In Fig. 2.30, we see the result of the construction of a triangle from one side and the two adjacent angles. The complete triangle is made by copying the given constituent side and two angles (the parameters) shown to the left of the triangle (using the two macros COPY-ANGLE and COPY-LENGTH). We can measure the three parameters, and the three sides and angles of the complete triangle (Fig. 2.31). We thus have available a graphic function with three input parameters and three output parameters. Variation of an input parameter (by dragging) alters one or more of the output parameters (e.g. from Fig. 2.31 to Fig. 2.32 to Fig. 2.33 to Fig. 2.34). If the two angles

40

are made to sum to 180°, the triangle degenerates since two sides are now parallel and one vertex moves off to infinity (Fig. 2.35).

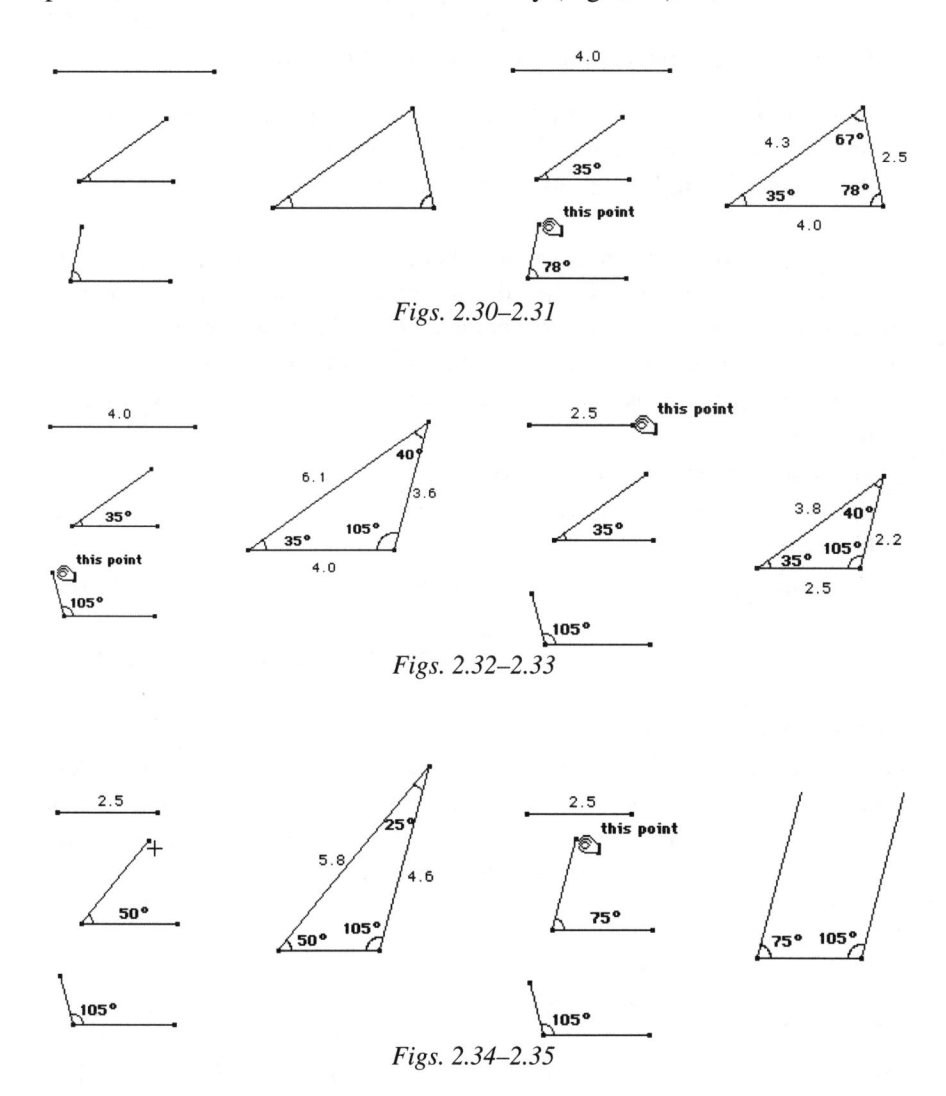

Figs. 2.30–2.31

Figs. 2.32–2.33

Figs. 2.34–2.35

2.5 CONSTRUCTING LOCI

The locus is the path followed by a dependent point related to some other independent point which is itself constrained to move on a predetermined path. The operative question for the teacher to pose is: "The point Y is constructed to be dependent on the point X. What path is described by Y when X is moved along a given path?"

Weaknesses in the conventional construction of loci are:

- they are very time-consuming and tedious to draw usually requiring repetition of the same construction many times over;
- they tend to be imprecise, involving free-hand interpolation.

Loci construction rarely features in school geometry – at most, it usually amounts to mentioning loci when discussing curves. Sometimes a function plotting program is used. Cabri-géomètre provides new opportunities for the interactive generation of loci. Suitable situations are:
- in the heuristic phase of solving construction tasks (i.e. exploratory investigational work);
- in experimental verification of construction results;
- in the construction of second order algebraic curves (conics), and also higher order curves;
- in examining the properties of special points in a triangle (e.g. centroid).

As illustrations we provide two examples from lessons involving the circle.

Example 8 *Pascal's limaçon*
Given a fixed point P and a fixed circle, we construct a tangent to the circle and a perpendicular to the tangent which passes through P (Fig. 2.36). Tangent and perpendicular meet at the point F. If we move the tangential point round the circle the point F generates Pascal's limaçon (Fig. 2.37).

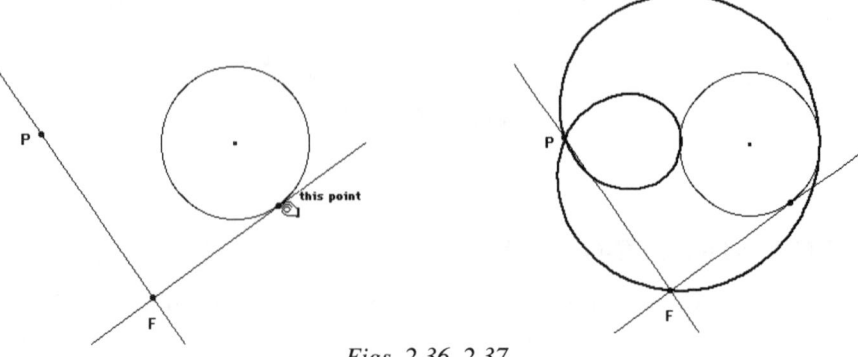

Figs. 2.36–2.37

Example 9 *Centroid and circle*

Which path is described by the centroid G of a triangle ABC, when one vertex (C) is moved along the triangle's circumcircle? (Figs. 2.38–2.40) The locus is shown as an emboldened circle in Fig. 2.40 and the circumcircle itself is hidden. (Note: when constructing this it may be helpful to start with a basic circle and define the triangle in terms of three points lying on the circle. This back-to-front approach avoids a potential problem of the circle changing size as C moves round it which can arise if the triangle is drawn first and the circumcircle constructed to pass through the three vertices.)

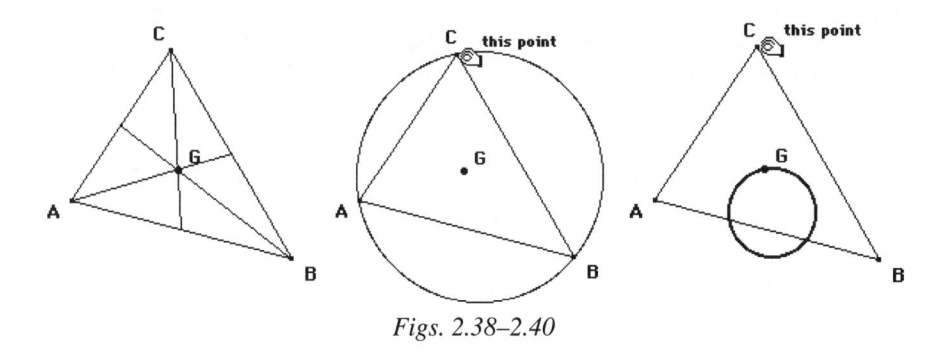

Figs. 2.38–2.40

2.6 DEFINING AND APPLYING MACRO-CONSTRUCTIONS

Basic constructions are often required as part of larger constructions. For example, it would be useful to have available as a menu option a construction for the centroid of a triangle. This could then be used in further constructions. The user would merely be required to supply the initial triangle. As a second example consider a typical drawing to illustrate Pythagoras' Theorem. This requires constructing a triangle and the three squares on the triangle's sides, and then possibly a further construction to show that the areas of the two smaller squares together equal that of the larger square. It could be useful to have all this available as a menu option with the user just supplying the triangle.

Weaknesses exist in conventional construction methods whether using ruler-and-compasses or a simple computer graphics package. The need to construct everything oneself has the following drawbacks:

– it discourages trial and error;
– it diverts attention away from the aim of the overall construction;

- it may significantly reduce the accuracy of the whole construction;
- it is extremely time-consuming, without producing new insight or understanding, more so because of forgetfulness or fatigue;
- it makes the final drawing appear complex and unclear because of the clutter of all the intermediate lines which must be drawn to achieve the final construction;
- it may lead to the actual construction process (many small steps) failing to correspond to the mental modular representation of the construction process (a few large steps).

Cabri-géomètre provides new possibilities with its facility to define macro-constructions as graphic functions. Fuller details are to be found in Chapter 3; here we briefly outline the principles. A macro-construction is made by:

- firstly constructing the drawing in the normal way;
- then selecting the **Macro** command from the **Miscellaneous** menu;
- then clicking on the desired initial object(s) which must be supplied to the macro;
- then clicking on the desired final object(s) which the macro is to produce;
- finally giving the macro a unique name.

It is important to note and remember what initial object(s) must be supplied for the macro!

The interactive use of a prepared macro requires opening the **Construction** menu and clicking on the name of the macro (in Chapter 1 an illustration was given in Figs. 1.30–1.34) and clicking on the initial object(s), which must already be on the screen. The final objects then are automatically produced by Cabri-géomètre and can be used in further constructions or investigations. In addition to defining the macro by direct clicking on the basic objects, the basic objects can also be taken from a previous figure or configuration and should be clicked on there. Normally the objects to be clicked on will lie *inside* the figure. However, it is possible to have basic objects *outside* the figure which depends upon them.

Example 10
In Fig. 2.41 P and Q are basic objects and a circle is drawn with centre Q passing through M the midpoint of PQ. Varying P causes M to move and consequently causes the circle to change.

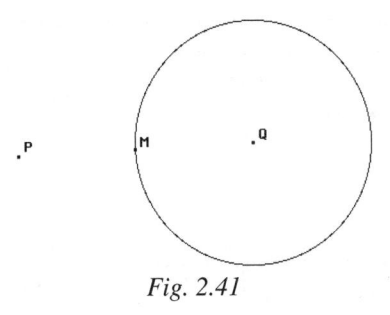

Learning geometry through interactive construction

Fig. 2.41

We illustrate this powerful and flexible tool by means of two examples.

Example 11 *Circumcircle of triangle*
Construction of the circumcircle of a triangle requires finding the intersection point of two perpendicular bisectors which provides the centre of the required circle (Fig. 2.42). The first step to creating the appropriate macro is to designate the initial object to be the triangle as an entity or alternatively designating the three vertices as the initial objects (Fig. 2.43, shown bold).

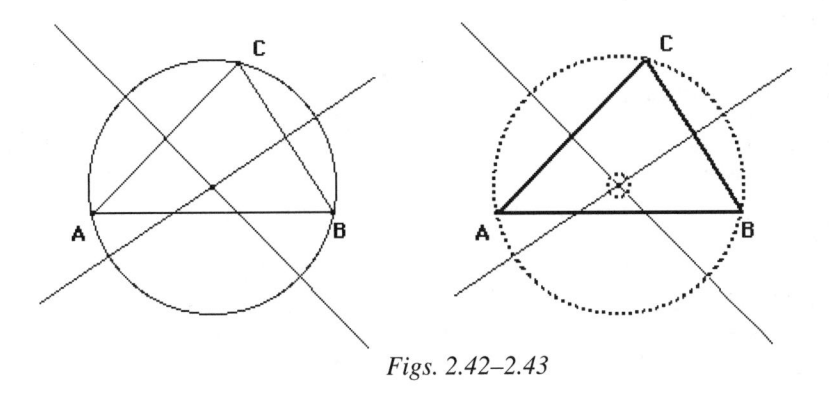

Figs. 2.42–2.43

Then the circumcentre and circumcircle are designated as the final objects (Fig. 2.43, shown dotted). Finally the macro-construction is named (Fig. 2.44).

Fig. 2.44

The name of the new macro appears at the end of the **Construction** menu and can be selected from there (Fig. 2.45). Given any triangle (Fig. 2.46) the macro will draw the desired final objects – the circumcentre and circumcircle – for use in further constructions (Fig. 2.47).

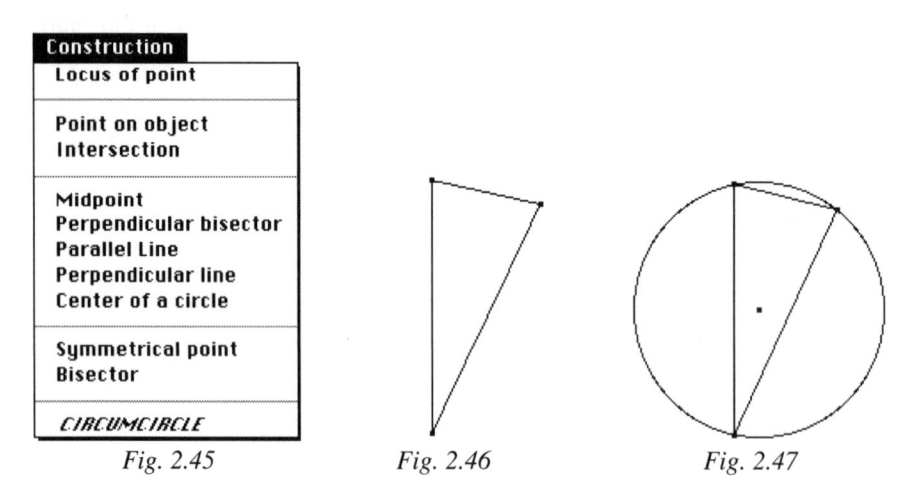

Fig. 2.45   *Fig. 2.46*   *Fig. 2.47*

Example 12 *Transformation of a quadrilateral into a square of equal area*
To solve this problem we modularise it by defining the three following macros, which are themselves based on normal constructions:

(1) transform a quadrilateral (defined by 4 points) into a kite of the same area;
(2) transform a kite (defined by 4 points) into a rhombus of the same area;
(3) transform a rhombus (defined by 3 points) into a square of the same area.

These macros when defined appear at the end of the **Construction** menu (Fig. 2.48). We utilise these macros for the total solution of the problem as follows:

(a) We apply the first macro [QUAD-KITE] to an initial quadrilateral ABCD (Fig. 2.49) and obtain a kite AB'CD' (Fig. 2.50);
(b) We apply the second macro [KITE-RHOMBUS] to the kite and obtain a rhombus A'B'C'D' (Fig. 2.51);
(c) We apply the third macro [RHOMBUS-SQUARE] to the rhombus to obtain the square A"B"C"D' (Fig. 2.52).
(d) Then we could define a fourth macro [QUAD-SQUARE] which transforms any general quadrilateral into a square of equal area, by combining the three macros described above (Fig. 2.53).

In all these macros the basic geometrical property employed is that all triangles on the same base with the same height are equal in area.

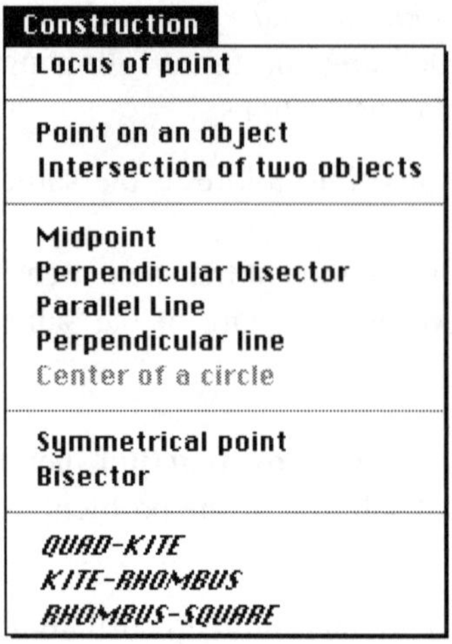

Discovering Geometry with a Computer

Construction
Locus of point

Point on an object
Intersection of two objects

Midpoint
Perpendicular bisector
Parallel Line
Perpendicular line
Center of a circle

Symmetrical point
Bisector

QUAD-KITE
KITE-RHOMBUS
RHOMBUS-SQUARE

Fig. 2.48

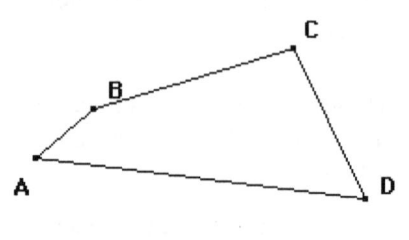

Fig 2.49

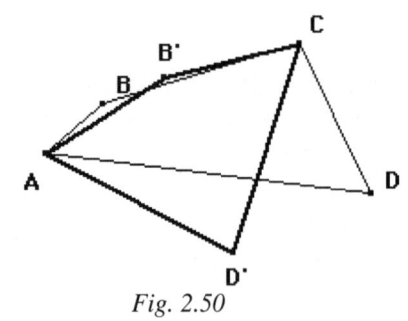

Fig. 2.50

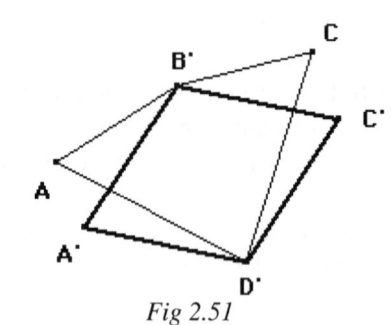

Fig 2.51

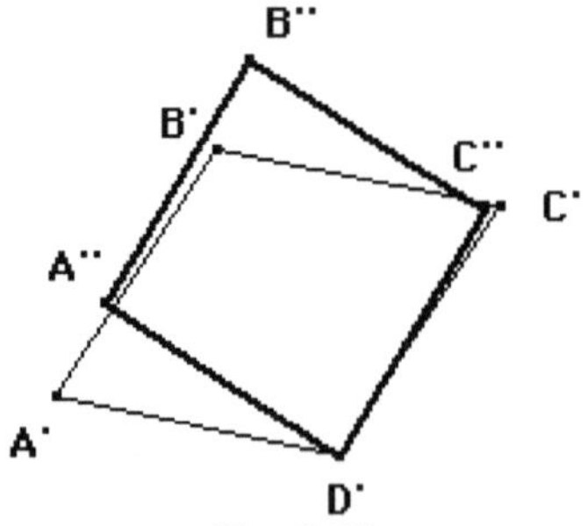

Fig. 2.52

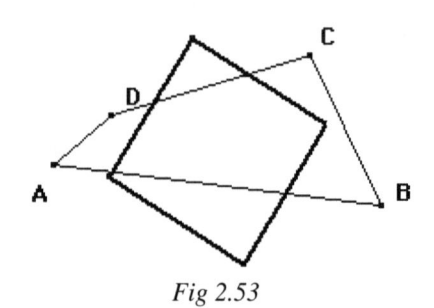

Fig 2.53

48

Example 13 *Richmond's construction of the regular pentagon inscribed in a circle*

To inscribe a regular pentagon ABCDE in a circle centre O, proceed as follows:

(1) draw a line segment OA in a circle centre O, radius OA,
(2) draw a line perpendicular to OA through O to intersect the circle at P,
(3) find M the midpoint of OP and draw in line segment AM,
(4) mark angle OMA and draw its bisector to obtain N where the bisector intersects OA,
(5) draw the perpendicular to OA through N to obtain B and E where it intersects the circle (this gives AB as a side of the pentagon as shown in Fig. 2.54).
(6) draw a circle centre B, radius BA to intersect the original circle at C,
(7) draw a circle centre E, radius EA to intersect the original circle at D,
(8) draw in the line segments for the five sides,
(9) finally, hide the auxiliary lines.

The implementation as a macro is left as an exercise for the reader.

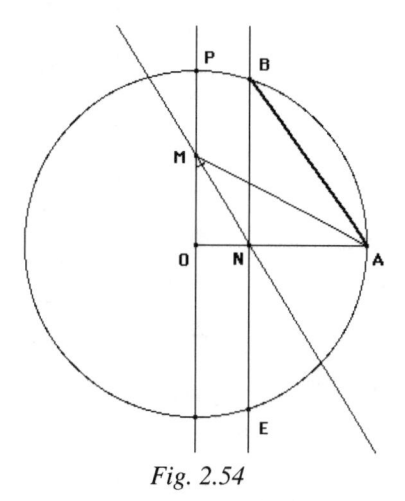

Fig. 2.54

Richmond provided a similar construction for the regular 17-sided polygon (see Coxeter, 1961 or Wells, 1991 for the details).

2.7 SOLVING CONSTRUCTION PROBLEMS

Weaknesses using conventional tools for solving construction problems are:

- there is limited support for the heuristic (exploratory) phase of the solving process;
- there is difficulty in correcting construction errors;
- there is no possibility of re-positioning or re-scaling the figure;
- there is no possibility of repeating the construction without going through all the steps again (this requires a macro-construction facility).

Cabri-géomètre is particularly useful in the heuristic phase of construction problem solving (see Diagram 2.2 below where an arrow means "... supports ...").

Varying figures
Measuring lines, angles, areas
Use of basic construction macros
Defining and using of constant macros
Generation of loci
Repetition of constructions
Use of further system functions
*Generating, erasing, inserting, cutting out, emphasizing, shifting, copying objects.
Running macros.*

Heuristic Phase
Algorithmic Phase
Analytical Phase

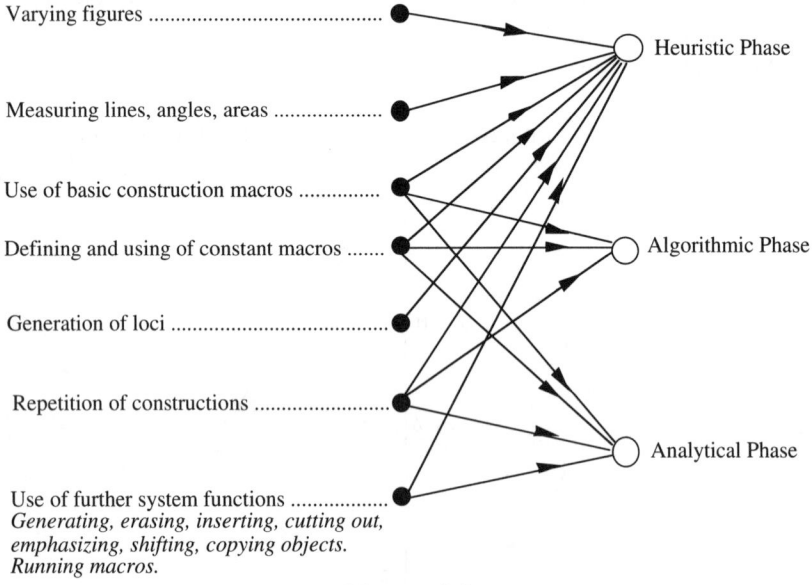

Diagram 2.2

The **heuristic phase** can be described by the following typical activities:

- drawing a configuration which satisfies the given conditions;
- varying the configuration paying particular attention to all special cases to obtain an overall picture of the situation;

- making initial observations about the situation (conditions for existence, inherent properties e.g. the sum of lengths two sides of a triangle must exceed the third, degenerative cases, ...);
- amending a configuration by drawing in auxiliary lines;
- seeing relationships in the configuration;
- using heuristic strategies:
    simplifying the problem by omitting some of the conditions
    working backwards from the solution
    breaking the problem into stages – modularising.

These activities can be well supported by Cabri-géomètre in drag-mode in seeking an experimental solution which satisfies the given conditions. This may involve interactive locus generation.

The **algorithmic phase** is characterised by the execution of the discovered method of solution utilising basic constructions whereby the solution is uniquely determined by the given objects. We note in passing that all constructions implicitly use uniqueness and existence theorems e.g. we naturally assume that there exists one unique perpendicular line from a point to a given straight line and do not normally make this explicit in geometrical work. This algorithmic phase concludes with the definition of a macro-construction. The **History** mode allows the construction process to be reviewed and written down.

Below we illustrate problem solving activities, particularly the heuristic phase, using the example of an *insertion* problem.

Example 14

Two straight lines a and b and a free point P are given (Fig. 2.55). We seek an equilateral triangle which has P as one vertex and whose two other vertices lie on a and b (Fig. 2.56).

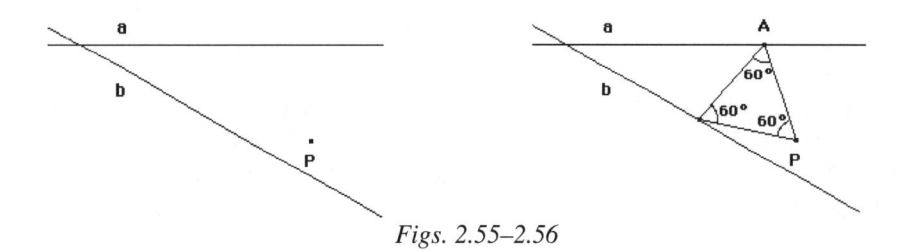

Figs. 2.55–2.56

Heuristic phase

We construct a point A, free to move along the line a, and an equilateral triangle with side PA. We could construct the triangle using normal geometrical techniques or, if available, use a macro (EQUILATERAL-TRIANGLE) which would require supplying two points (first P then A, say) representing two vertices of the triangle. We label the third vertex B (Fig. 2.57: heuristic method – leaving out a condition), P, A, B forming an *anti-clockwise* triplet. Variation of the position of point A on line a generates an experimental solution with point B on line b (Fig. 2.58).

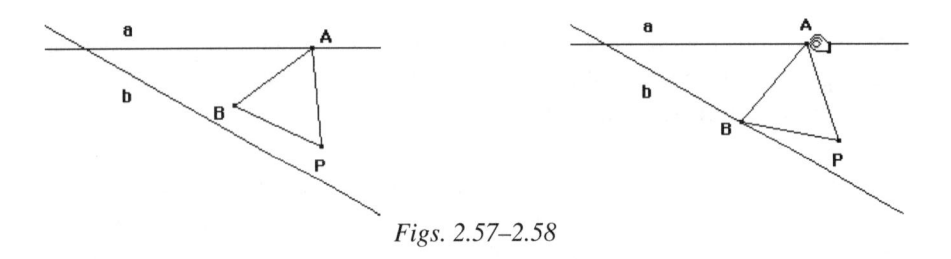

Figs. 2.57–2.58

This is the first resolution of the task. (Note: Varying the position of P is an alternative way to approach such a solution but is not pursued here.) We get another experimental solution by applying the macro EQUILATERAL-TRIANGLE to A, P instead of to P, A; in which case P, A, B' forms a *clockwise* triplet (Fig. 2.59–2.60).

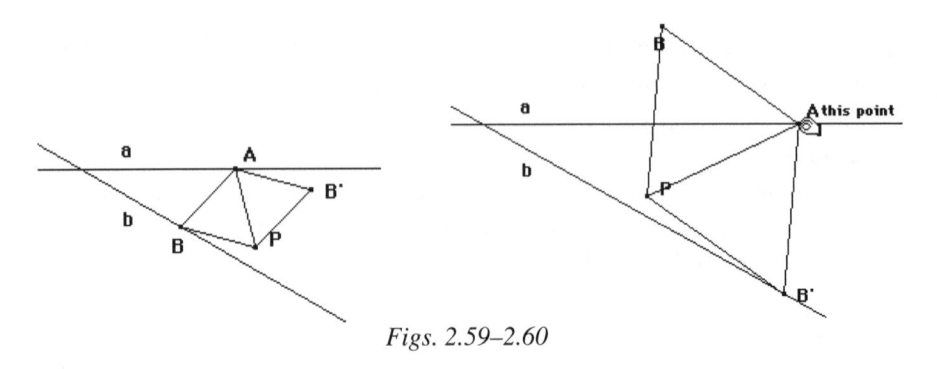

Figs. 2.59–2.60

As A moves along a, leading to an experimental solution, B describes a locus whose intersection with b determines the sought for third vertex (Fig. 2.61–2.62: locus method; in this dynamic process as A moves the equilateral triangle rotates around P and changes size).

Figs. 2.61–2.62

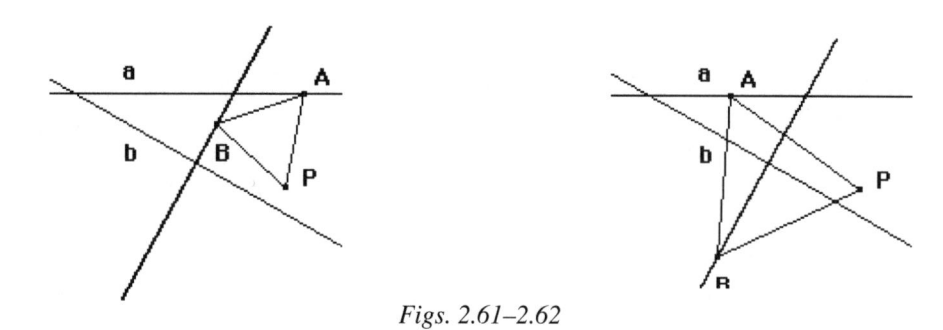

The locus generated using the **Locus** command is found to be a straight line. Unfortunately this 'working backwards' method is not fully practicable since nothing more can be constructed with the straight line locus produced in Fig. 2.62. (A locus in Cabri-géomètre is just a series of points on the screen and has no internal logical representation. This is an unfortunate, but understandable, shortcoming of the current version.) We try to overcome this weakness by rotating line a around P until the image line a' passes through B (Figs. 2.63–2.65). To check, A is moved along a, and B must meanwhile run along a'. At the same time, we can also use the **Locus** command (the locus of B must then coincide with a). Measuring shows the angle between a and a' to be 60°; thus a must be rotated 60° about P (Fig. 2.66). That completes the heuristic phase of the solution search.

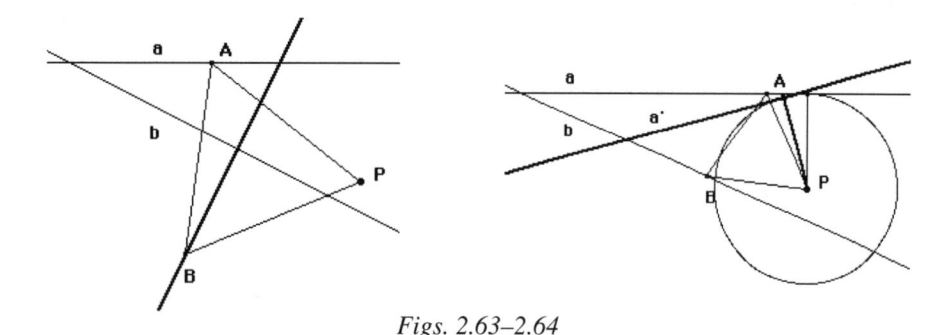

Figs. 2.63–2.64

Figs. 2.65–2.66

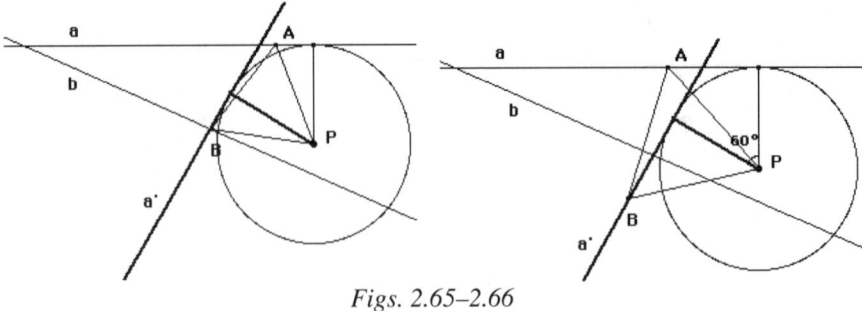

Algorithmic phase

In the **History** mode we can see the detailed construction with all intermediate lines (Fig. 2.67 shows **History** applied to Fig. 2.66 with the very last construction step being reported).

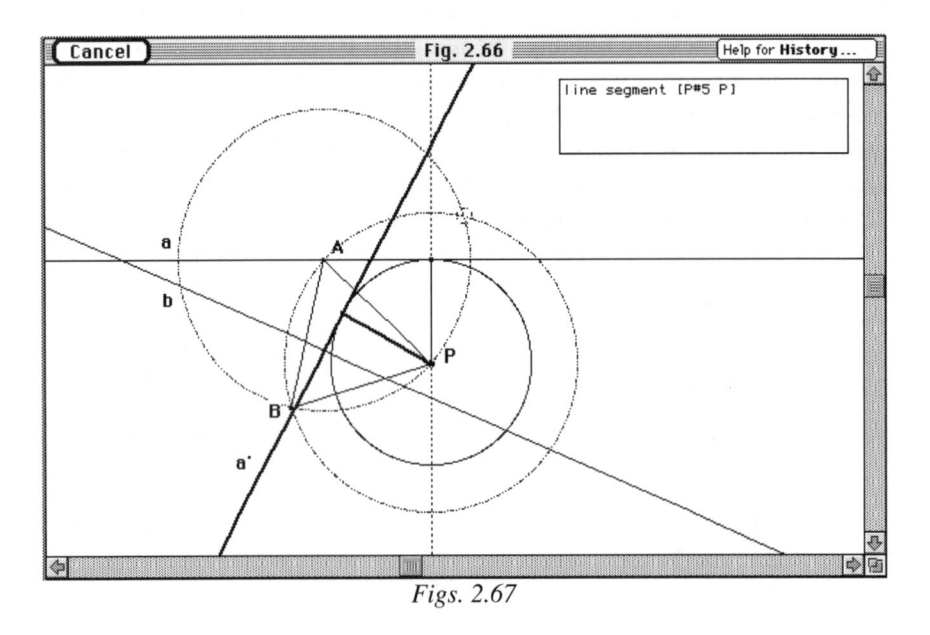

Figs. 2.67

We conclude the algorithmic phase by defining a macro with initial objects a, b and P (shown bold in Fig. 2.68) and the final objects (shown as dotted lines and dotted circles in Fig. 2.68). We call the macro INSERT-TRI-ANGLE (Fig. 2.69). We apply it to any point-and-two-line configuration (Fig. 2.70), doing so twice (with the order of selection of the two lines reversed on the second occasion) to get both solutions (Fig. 2.71) and vary

54

the positions of the lines and point, in order to test the macro. (Note: the orientations of the lines can be altered by holding down the SHIFT key while dragging.) A systematic investigation could follow, examining the solubility conditions in terms of the angle between the straight lines and the position of P.

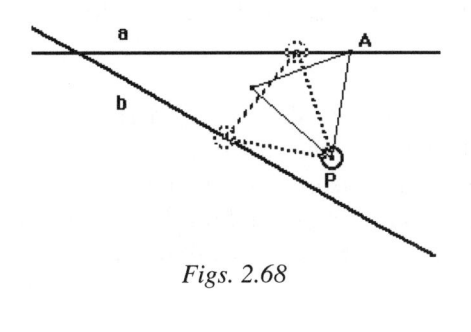

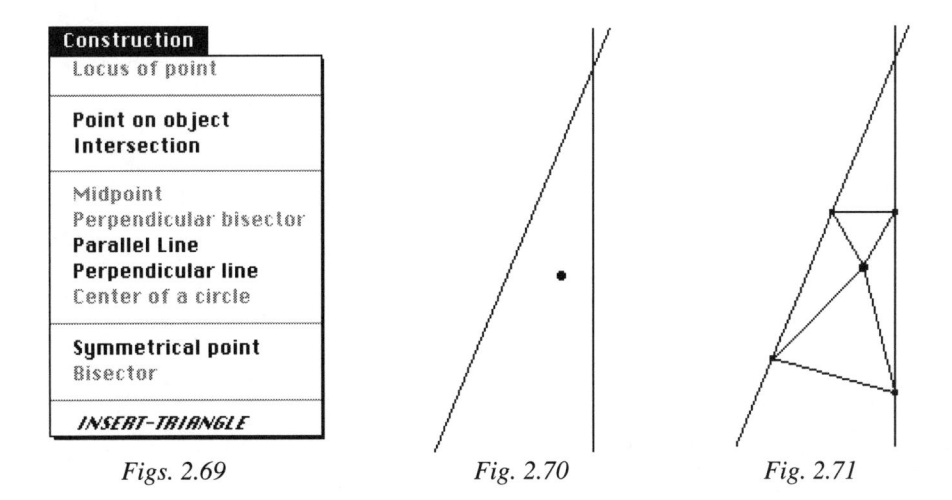

Figs. 2.68

Figs. 2.69   *Fig. 2.70*   *Fig. 2.71*

2.9 FINAL REMARKS

14 year old students who have used Cabri-géomètre in school as a construction tool have displayed a high level of motivation to use the system; experience shows that the local **Help** facility simplifies its initial use. This positive experience has been found for all those uses outlined in sections 2.2.1–2.2.5. However, it should be noted that when solving construction problems

students seem content with the experimental solution which they obtain by using drag-mode! It is helpful if students are advised that:

- a correct construction will be stable in drag-mode when the initial objects are varied;
- the solution is to be completed with the definition of a macro-construction;
- the corresponding compasses and ruler construction should be studied also.

Our experience with students shows that mastery of ruler-and-compasses constructions (either performed practically or using a computer drawing package) is an important learning prerequisite to being able to build a repertoire of macro-constructions and for discovering geometric properties by studying invariance of objects under drag-mode transformation.

CHAPTER 3

Creating macros for basic geometrical constructions

*When we build, let us think
that we build for ever.*
                John Ruskin

3.1 INTRODUCTION

The following basic constructions are often met in elementary geometry:

- construction of a perpendicular bisector
- bisection of a straight line segment (i.e. construction of midpoint)
- bisection of an angle (i.e. construction of a line bisecting an angle)
- erection of a perpendicular on a line
- dropping of a perpendicular from a point to a line
- construction of a parallel line
- copying a line segment (replicating a length)
- copying an angle (replicating an angle)

The following elementary operations are used in carrying out these basic constructions using ruler and compasses:

- placement of points freely in the plane or on straight lines or circles
- construction of a straight line passing through two points
- construction of a circle about a centre passing through a given point
- construction of the point of intersection of two straight lines
- construction of the points of intersection of two circles
- construction of the points of intersection of a straight line and a circle

These basic constructions are building blocks (modules) used in more complex constructions, and can be used as a short-hand way of describing or defining complex constructions. They help in modular working and thinking in geometry. Naturally, computer tools for geometric constructions have permanently available system macros for such basic constructions and the use of these macros facilitates construction. Cabri-géomètre, for example, has system macros for all the above basic constructions except the two copying constructions.

Such *system* macros are 'black boxes' i.e. the student as user has no insight into their definitions. This is in contrast to *user-defined* macros, of course, where the user has an intimate knowledge of the definition. For this reason we develop in this chapter possible definitions for the basic constructions which simulate elementary ruler-and-compasses constructions. We start with a set-up for a beginner of lower secondary school age for whom the **Creation** and **Construction** menus have been reduced by using the **Edit menus** customisation facility located in the **Miscellaneous** menu (see Fig. 1.15 in Chapter 1 for the layout of the **Miscellaneous** menu). Figs. 3.1 and 3.2 show the customised menus which are to be produced.

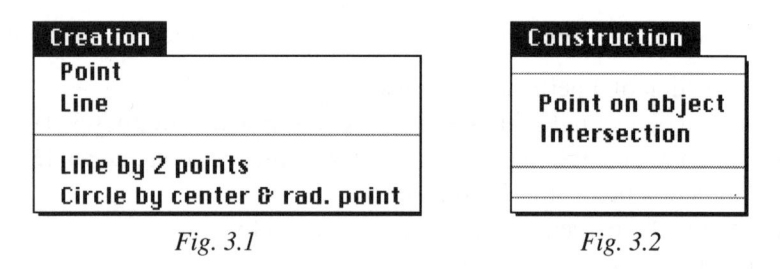

```
┌─Creation──────────────────┐
│ Point                     │
│ Line                      │
├───────────────────────────┤
│ Line by 2 points          │
│ Circle by center & rad. point │
└───────────────────────────┘
```
Fig. 3.1

```
┌─Construction──────────────┐
│ Point on object           │
│ Intersection              │
├───────────────────────────┤
└───────────────────────────┘
```
Fig. 3.2

In this exercise it is the task of the students to enlarge the **Construction** menu by defining macros for the basic constructions. This is to include some which are not provided with the system but which are very useful e.g. the copying of line segments and copying of angles. These will be put to good use later in this book. Because Cabri-géomètre only allows sequential constructions to be carried out in user-defined macros we must expect differences between students' and Cabri-géomètre's own system macros for basic constructions with regard to the choice of initial (basic) objects and also in their general validity.

3.2 USER-DEFINED MACROS FOR BASIC CONSTRUCTIONS

In order to complete basic construction tasks we need to preserve only those objects shown in the reduced **Creation** menu (Fig. 3.1) and the

Creating macros for basic geometrical constructions

actions shown in the reduced **Construction** menu (Fig 3.2), and, of course, the option of defining macros! In this chapter, where appropriate, we use our own user-defined macros to solve further basic construction tasks, building up a system in the manner familiar to LOGO programmers.

In order to create flexible macros, only points are used as initial objects. In the definition of a macro a short piece of text may be added to explain the macro – which can serve as on-line help. The sequence of basic construction tasks set out in this chapter has been found effective, although variations are possible. The most generally valid solution is always given, although there may be others.

3.2.1 Construction of the perpendicular bisector and the midpoint of a line segment

We are given two distinct points A and B as end-points of a line segment (Fig. 3.3). We create, in the manner of Euclid, the two circles with centres A and B and common radius [AB], then generate the two points of intersection of the circles (C and D), and link C and D with a straight line. The line CD intersects AB at M. The required perpendicular bisector of AB is the line though C, D and the required midpoint of AB is M.

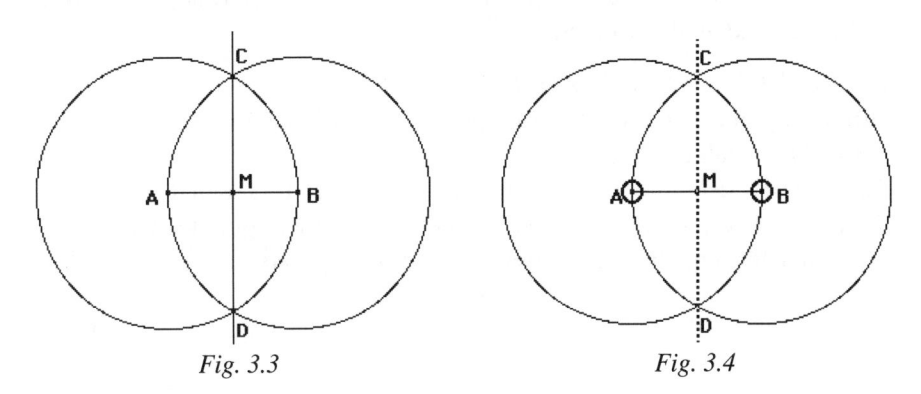

Fig. 3.3   *Fig. 3.4*

Macro PERP-BISECTOR (Fig. 3.4)
Initial objects: Two points (A, B – marked by small circles).
Target object: Perpendicular bisector (through C, D – marked as a dotted line).
Text: "Click on two points determining the line segment whose perpendicular bisector is required."

59

Discovering Geometry with a Computer

Macro MIDPOINT (Fig. 3.5)
Initial objects: Two points (A, B – marked by small circles).
Target object: Midpoint (M – marked as a small dotted circle).
Text: "Click on two points determining the line segment whose midpoint is required."

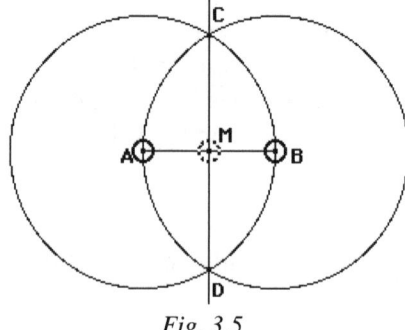

Fig. 3.5

3.2.2 Construction of the angle bisector

We reformulate the construction of the angle bisector of angle ACB (Fig. 3.6) as the construction of a perpendicular bisector. To do this, we construct a circle with centre C and radius [CB] which intersects the line through C, A at D. The line produced by the macro PERP-BISECTOR applied to B, D is the required angle bisector.

Macro ANGLE-BISECTOR (Fig. 3.7)
Initial objects: Three points (side point A, vertex point C, side point B, – marked by small circles).
Target object: Angle bisector (through C – marked as a dotted line).
Text: "Click successively on a side point, then vertex point, then other side point to determine the angle whose bisector is required."

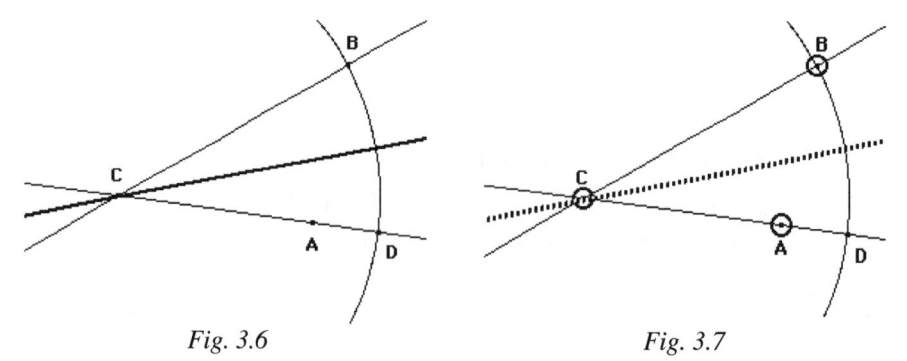

Fig. 3.6   *Fig. 3.7*

3.2.3 Construction of a perpendicular

Here again we reformulate the construction of a perpendicular through a point as the construction of a perpendicular bisector. We are given a straight line passing through points A and B (or a line segment AB) and on it a point C which is generated with **Point on object** (Fig. 3.8). We construct a circle centre C with radius [CA] which intersects the line passing through A and B at point D. Using the macro PERP-BISECTOR applied to A, D we obtain the perpendicular to the straight line passing through C.

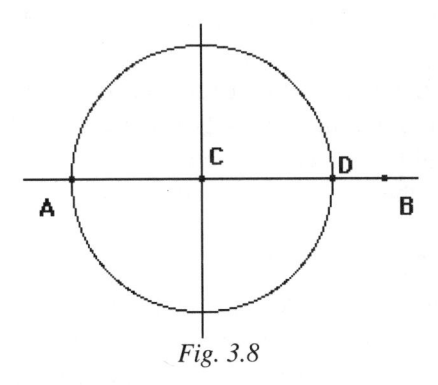

Fig. 3.8

Macro PERP-ERECT (Fig. 3.9)
Initial objects: Three points (A and B defining a line or line segment, and C lying on the line or line segment – marked by small circles).
Target object: Perpendicular passing through a specific point (C) on a line or line segment (– marked as a dotted line).
Text: "Click on two points determining the line or line segment, then click on a third point also on the line or line segment, through which the perpendicular line is to pass."

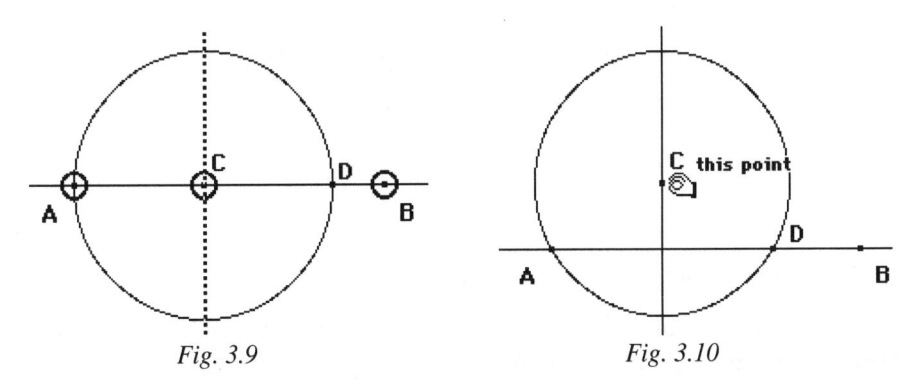

Fig. 3.9   *Fig. 3.10*

Note: With this macro it is impossible to construct the perpendicular passing through either of the points which determine the line (i.e. A and B) because the macro requires three **distinct** points as initial objects. (We can only define macros on the basis of sequential constructions and cannot have branching to handle different cases.) This limitation does not apply to the system macros which are supplied. Furthermore, if we define the macro utilising a line segment instead of two basic points as the initial object, the macro cannot construct the perpendicular through one of the points delineating the line segment (for the reasons outlined above) although it can be done through the other (which is which cannot be predicted).

We can generalise the construction of the perpendicular by releasing point C from its position on the line through A and B (Fig. 3.10). Then the construction will be valid for erecting a perpendicular as well as for dropping a perpendicular (with the above-mentioned limitation).

Macro PERP-CONSTRUCT (Fig. 3.11)
Initial objects: Three points (A and B defining a line or line segment, and C lying on or outside the line or line segment – marked by small circles).
Target object: Perpendicular passing through a specific point on or outside a line or line segment (– marked as a dotted line).
Text: "Click on two points determining the line or line segment, then click on the point through which the perpendicular line is to pass."

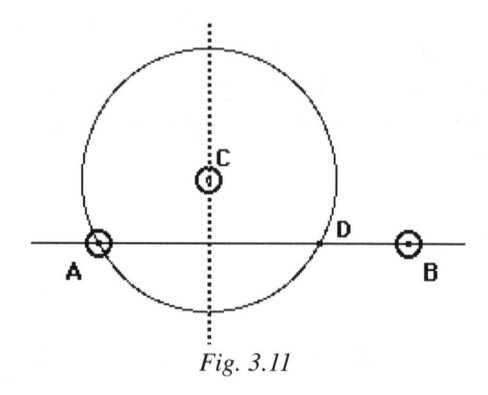

Fig. 3.11

3.2.4 Construction of a parallel line

We are given a straight line, or line segment, through two points A and B (Fig. 3.12) and a point C which does not lie on the line. In order to construct the parallel to AB passing through C we choose a construction which uses the macro PERP-CONSTRUCT.

To this end, we drop the perpendicular from C onto AB obtaining the 'foot' D. A circle centre D with radius [DC] intersects CD extended at E. Now we can erect the perpendicular at C to the line DE, which is the required line parallel to AB.

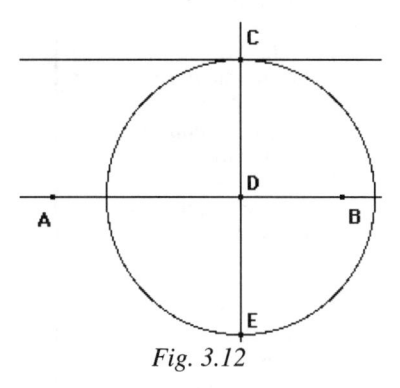

Fig. 3.12

Macro PARALLEL (Fig. 3.13)
Initial objects: Three points (A and B defining a line or line segment, and C lying outside the line or line segment – marked by small circles).
Target object: Parallel to a line or line segment passing through a specific point outside the line or line segment (– marked as a dotted line).
Text: "Click on two points determining the line or line segment, then click on a third point (lying outside the line or line segment) through which the parallel is to pass."

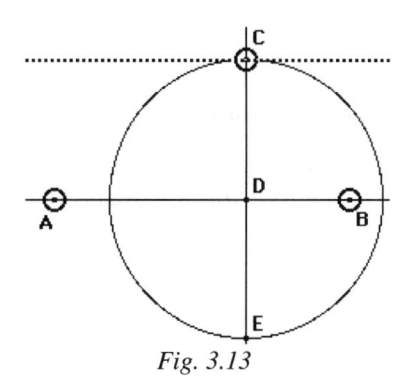

Fig. 3.13

Extended Construction menu
By adding these macros we obtain the extended **Construction** menu shown in Fig. 3.14.

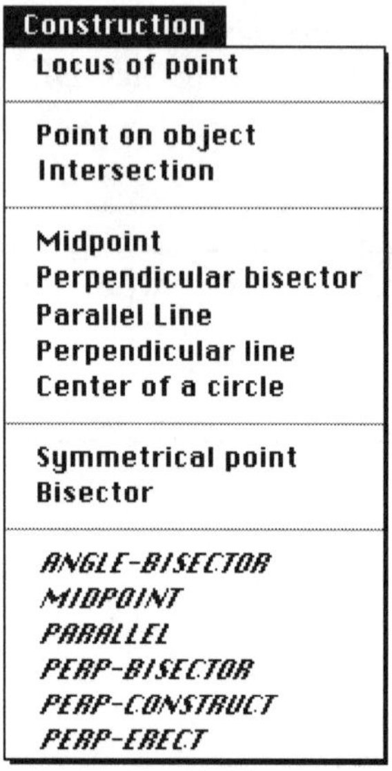

```
Construction
 Locus of point

 Point on object
 Intersection

 Midpoint
 Perpendicular bisector
 Parallel Line
 Perpendicular line
 Center of a circle

 Symmetrical point
 Bisector

 ANGLE-BISECTOR
 MIDPOINT
 PARALLEL
 PERP-BISECTOR
 PERP-CONSTRUCT
 PERP-ERECT
```

Fig. 3.14

We still lack macros for the transferring (or copying) of a line segment and of an angle. First we develop two macros for the reflection of a point about a point and for the reflection of a point in a straight line, which will be needed for the two copying macros.

Similar macros can be developed for hyperbolic geometry as discussed in Chapter 11 section 11. 5.

3.2.5 Construction of a point-symmetric point

We are given a point A and a point B about which A is to be reflected (Fig. 3.15). We construct a straight line through A and B and then generate a circle centre B radius [BA] which intersects the straight line at the required image point A'.

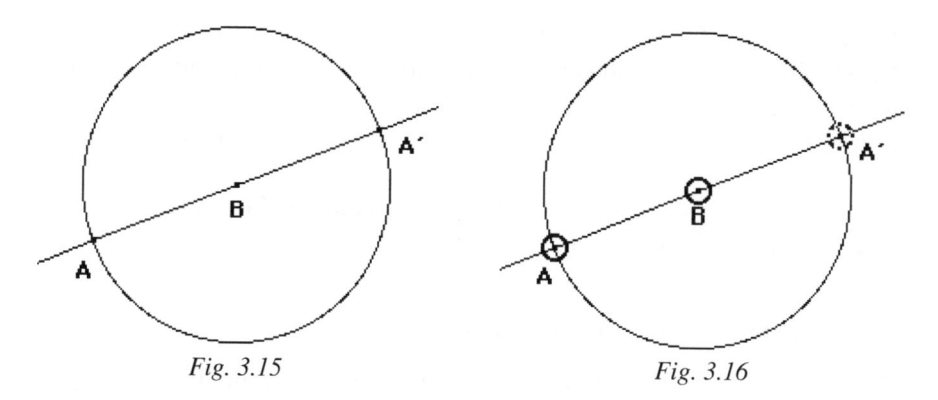

Creating macros for basic geometrical constructions

Fig. 3.15

Fig. 3.16

Macro PT-SYMMETRIC-PT (Fig. 3.16)

Initial objects: Two points (A as given point and B as centre for reflection – marked by small circles).

Target object: Image point A' point-symmetric to A (– marked as a small dotted circle).

Text: "Click on the point which is to be reflected, then click on a the point which is the centre of reflection."

3.2.6 Construction of a circle with given centre and radius

As an application of the PT-SYMMETRIC-PT macro we now construct for a given line segment AB and a given point C the circle with centre C and radius [AB] (Fig. 3.17). To do this we find by macro MIDPOINT the midpoint D of A and C. C is then point-symmetric to A in centre D (so C = A'). Using the macro PT-SYMMETRIC-PT the image of B in D is produced (B'). The circle centre C passing through B' is the required circle because [A'B'] = [AB].

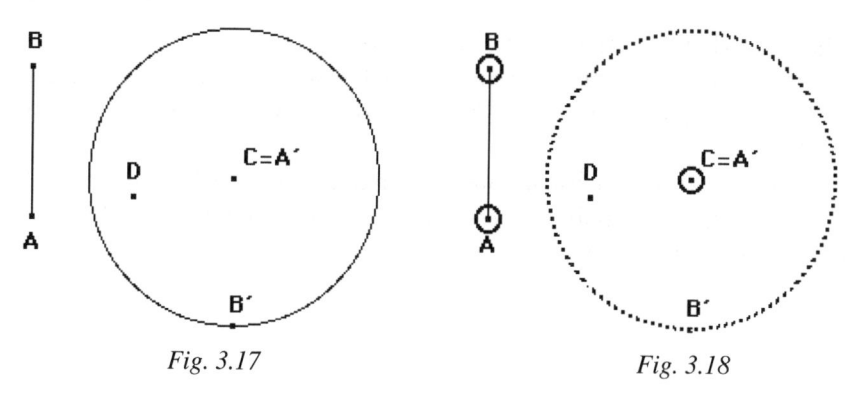

Fig. 3.17

Fig. 3.18

65

Macro CIRCLE-C-r (Fig. 3.18)
Initial objects: Three points (A and B defining the line segment, and a third point C – marked by small circles).
Target object: Circle with given radius and centre (– marked as a dotted circle).
Text: "Click on two points determining the radius of the circle, then click on a point determining the centre."

3.2.7 Construction of a line-symmetric point

We are given two points A and B and a point C not on the line through A and B. We seek the image of C by reflection in the line through A, B. To achieve this we use the so-called two-circle figure: i.e. circle centre A with radius [AC] and circle centre B with radius [BC] (Fig. 3.19). The second point of intersection of the two circles is C' which is symmetrical to C. In the special case when C lies on AB, then C=C' is the common tangent point of the two circles.

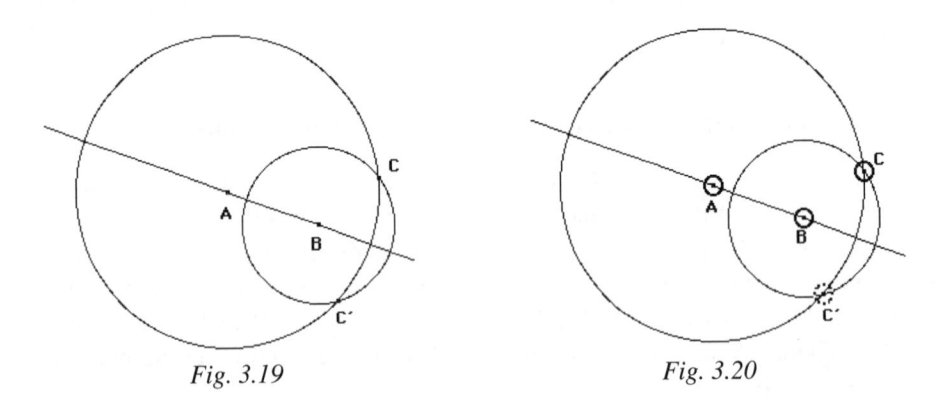

Fig. 3.19   *Fig. 3.20*

Macro LINE-SYMMETRIC-PT (Fig. 3.20)
Initial objects: Two points (A and B which determine the axis of symmetry and a further point C which is the point to be reflected – marked by small circles).
Target object: Symmetrical point C' (– marked as a small dotted circle).
Text: "Click on two points determining the reflection line, then click on the point which is to be reflected."

3.3 CONSTRUCTION OF COPYING MACROS

So far, the macros have been relatively easy to construct. We now turn to the more difficult copying problems, probably not understandable by most students, but included here because of their considerable utility and their absence from Cabri-géomètre itself.

3.3.1 Construction of a copy of a line segment (transferring a line segment)

We are given two points A and B defining a line segment and a straight line or line segment through two further points C and D. The line segment AB is to be copied so that it starts at C and lies along the line through C, D (Fig. 3.21).

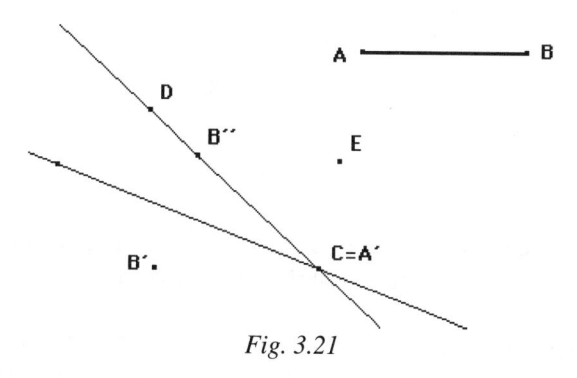

Fig. 3.21

In order to construct the second end-point of the copied line segment we use the four macros

MIDPOINT,
PT-SYMMETRIC-PT,
ANGLE-BISECTOR,
LINE-SYMMETRIC-PT.

- We use MIDPOINT to obtain E as the midpoint of [AC].
- We use PT-SYMMETRIC-PT to reflect B in E to give B'.
- We use ANGLE-BISECTOR to obtain the angle-bisector of B'CD.
- We reflect B' in the angle-bisector of B'CD using LINE-SYMMETRIC-PT to obtain B", which is the required end-point. [CB"] equals [AB].

Macro COPY-LENGTH (Fig. 3.22)

Initial objects: Four points (two points A and B as end-points of a line segment, two further points C and D which determine a line segment or straight line on which the copy is to lie – marked by small circles).

Target object: Second end-point of the copied line segment (– marked as a small dotted circle).

Text: "Click on the two points of the line segment to be copied, then click on a new end-point for the line segment copy, and on a further point which determines the direction in which the copied line segment must lie."

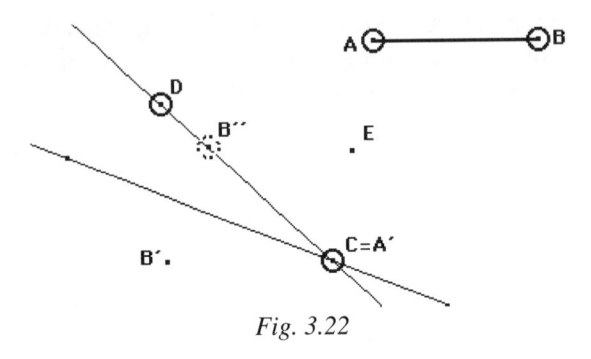

Fig. 3.22

Note 1: Where necessary, the found end-point can be reflected in the given end-point to reverse the direction of the line segment.

Note 2: An alternative way is to use the macro CIRCLE-C-r, defined in section 3.2.6 (Fig. 3.23). The initial objects (shown bold in Fig. 3.24) and the target object (shown dotted in Fig. 3.24) are the same as those stated immediately above. This is left to the reader to investigate.

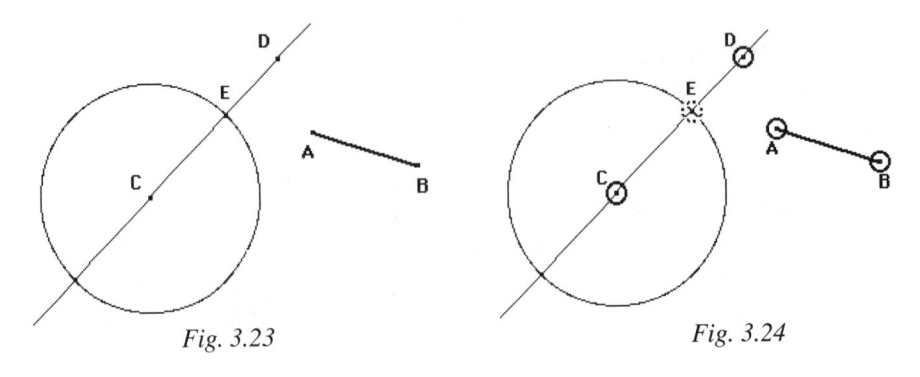

Fig. 3.23   Fig. 3.24

3.3.2 Construction of an angle copy (transferring an angle)

We are given an angle ACB and two other points D and E. The angle is to be copied to lie on DE (Fig. 3.25) where D is the new vertex and E is a side point of the angle in its new position. One of the four possible positions of the second side point will be constructed (the three others can be obtained by point or line reflection). We use in succession the following macros:

- MIDPOINT (initial objects C, D; target object F)
- PT-SYMMETRIC-PT (initial objects A (then repeated with B) and F; target object A' (then B'))
- ANGLE-BISECTOR (initial objects: A', D, E; target object: bisector of angle A'DE)
- LINE-SYMMETRIC-PT (initial objects: D and an auxiliary point on the angle bisector to specify the line, and A' as the point to be reflected (then repeated with B'); target objects: A" (and then B")).

A"DB" is the required copied angle.

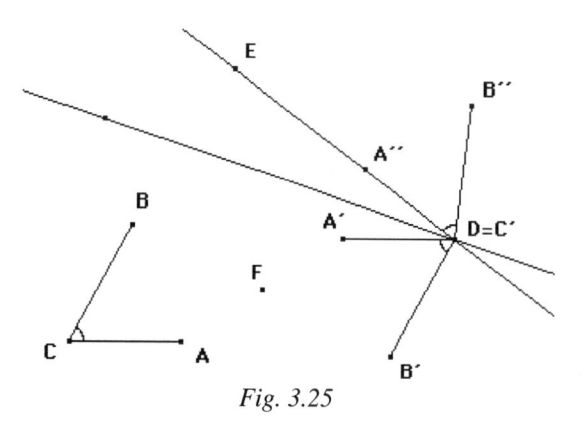

Fig. 3.25

Macro COPY-ANGLE (Fig. 3.26)
Initial objects: Five points (side point A, vertex C, side point B, new vertex D, and second line point E – marked by small circles).
Target object: Second side point of the copied angle (B" – marked as a small dotted circle).
Text: "Click on side point, vertex, and side point of the angle to be copied, then click on the new vertex and any other point on the line on which a side point of the copied angle is to lie."

Fig. 3.26

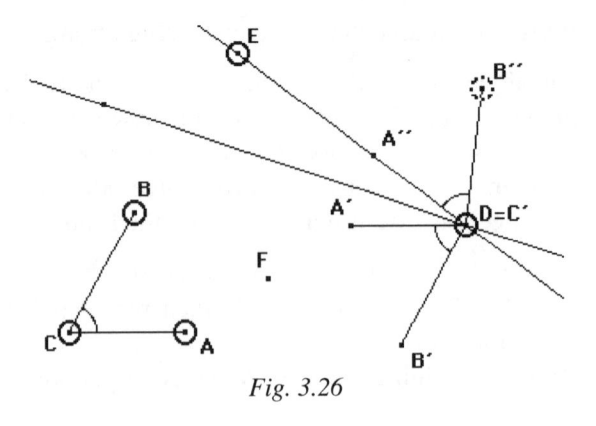

Alternative

The idea is to construct from angle ACB the isosceles triangle ACF and copy it using the macro CIRCLE-C-r so that one leg lies on the given line DE. Fig. 3.27 shows the result; Fig. 3.28 shows the five initial objects and the target object. This is left to the reader to explore.

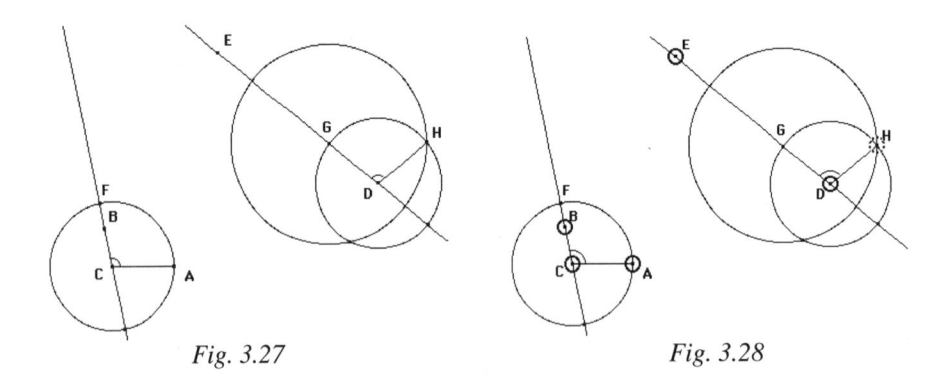

Fig. 3.27   *Fig. 3.28*

3.3.3 Extended Construction menu

By adding these macros we obtain the extended **Construction** menu shown in Fig. 3.29. Additionally, the macro CIRCLE-C-r could be included.

Creating macros for basic geometrical constructions

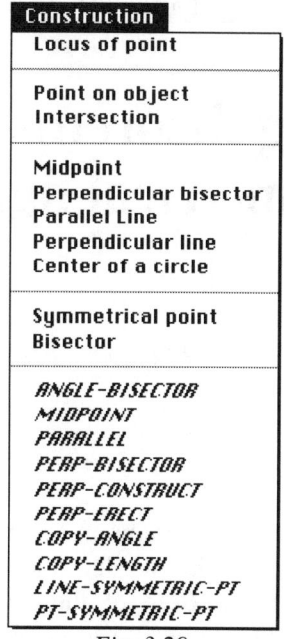

Construction
Locus of point
Point on object Intersection
Midpoint Perpendicular bisector Parallel Line Perpendicular line Center of a circle
Symmetrical point Bisector
ANGLE-BISECTOR *MIDPOINT* *PARALLEL* *PERP-BISECTOR* *PERP-CONSTRUCT* *PERP-ERECT* *COPY-ANGLE* *COPY-LENGTH* *LINE-SYMMETRIC-PT* *PT-SYMMETRIC-PT*

Fig. 3.29

- Students can compare the system macros with the user-defined macros which perform similar functions.
- The definition of a macro equivalent to **Centre of a circle** is possible in some versions of Cabri-géomètre only.

3.4 CONCLUDING REMARKS

The given constructions assume ruler and compasses to be the only tools. Cabri-géomètre is not suitable for the simulation of basic constructions using a right-angle, measuring ruler and protractor (sometimes combined in a set-square). This type of construction is modelled in the *Geometry Inventor* system.

The system of macros developed in Section 2 of this chapter may seem unnatural or unwieldy as compared to the normal ruler-and-compasses constructions. For example, only points are permitted as initial basic objects (rather than points, lines and circles) because some versions of Cabri-géomètre do not permit arbitrary points on objects to be used as basic objects for further constructions. Research is currently under way to eliminate such limitations.

CHAPTER 4

Discovering theorems by varying geometrical figures

Plus ça change, plus c'est la même chose.
Alphonse Karr

4.1 INTRODUCTION

The learning of mathematical theorems in secondary school mathematics has the two typical aspects of knowledge finding (induction) and knowledge confirming (deduction). Both of these overlapping concepts are important in mathematics teaching. Basic geometry, especially 2-D, is a suitable and traditional field for heuristic work because of the "small difference" between the ideal geometrical objects and their iconic models. For example, a drawn circle is clearly very similar to an idealised circle. The method of guided discovery of geometrical theorems is a useful teaching method; the teacher takes on a role of great importance as the organiser of the heuristic learning process.

A large proportion of the theorems in 2-D elementary geometry can be developed within the following scheme. (Of course it is difficult and not entirely satisfactory to force such a complex process into a linear scheme.)

Inductive theorem finding (through graphical construction)

Step 1: Solving a construction task.

Result: Geometric configuration (by a geometric configuration we mean here the combination of points, straight lines, semi-infinite lines, curves, circles, arcs, etc. with their incidences and relationships).

Step 2: Analysing the construction result (also by drawing in more details and emphasizing important elements, by measuring and by calculating on the basis of measurements).

Result: A first conjecture.

Step 3: Checking the conjecture with further examples which are to be constructed, with attention to differences between individual cases.

Result: Substantiation of the conjecture with the formulation of a (first) theorem or alternatively dismissal of the assumption as unfounded.

The weaknesses of the inductive theorem-finding process using conventional tools were mentioned in Chapter 2, and are elaborated upon here:

(1) It is time-consuming to prepare a sufficiently large number of representatives of the configuration relevant to the theorem under investigation. Constructions are often inaccurate which adds to the difficulty.
(2) It is time-consuming to take measurements from the configurations and to perform necessary calculations (even with a calculator). Measuring is subject to imprecision which complicates matters.
(3) Only static constructions result. These can only be made 'movable' in the imagination; we start out here from the *Pictorialistic hypothesis* which holds that mental ideas of figure transformations are formed analogously to the real transformations of the corresponding figures.

Using suitable software the computer can be used to help in specific cases, though in doing so one encounters a multitude of problems, only some of which we will deal with here.

The first two weaknesses above could be alleviated using sophisticated computer graphic tools for defining graphic procedures. By varying the values or positions of the numerical or graphic input variables which determine the configuration, one can generate a multitude of configurations. Where such graphics procedures are to be designed by the students themselves, a programming problem occurs: the programming is more difficult than the actual construction problem! This is so even in a user-friendly programming environment, and distracts from the aim of the activity which is the theorem finding. In some cases one must even know the relevant theorem to be able to do the programming! This makes the idea of discovering the theorem using the computer a nonsense.

In the case where graphic procedures are used as 'black-boxes', the student totally loses reference to the formation of the configuration, which is important in theorem-finding and proof-finding. In this respect, particular menu-controlled graphic systems for geometrical construction can help to establish a compromise between the different possibilities of using a computer to produce geometrical configurations and students themselves using conventional tools. Such graphics systems must possess menu options for basic constructions or have facilities for interactively defining macro-constructions. The student designs a configuration in interaction with the system, makes the system 'learn' the construction command sequence, and, finally, varies the objects which determine the configuration in order to produce further examples to provide an inductive base for theorem finding. Thus, weaknesses 1 and 2 can be countered; but not, however, weakness 3.

In the past, children's experience of animation was largely restricted to cartoon films. Now, with the use of graphic packages, the situation is changing. In programming environments currently available one can use recursive graphics procedures to generate geometrical cartoons, but the above reservations concerning programming problems and the use of black-box programs remain valid. However, Cabri-géomètre, is a further step forward. The changing of a configuration generally takes place through the student or teacher controlling a graphics cursor using a mouse. In this process, the screen is interpreted as an empirical model of a part of the Euclidean plane. Moreover, it is a drawing-sheet usable like a sheet of paper but drawn on in a special way. The representation of geometrical objects (e.g. circle, straight line) are of acceptable quality when using a well-adjusted high resolution screen, and compare satisfactorily with pencil-and-paper representation. The transition from one state to another takes place cinematically (in real-time) and for this to be truly convincing requires good hardware. The constant variation of the configuration can be combined with simultaneous measuring of lengths, angles and areas. The recording of a sequence of phase-pictures ('snapshots') of positions (by means of the **Save** command) allows the later review of initial, intermediate and final states of the configuration. The continuous modification process is always reversible.

The fundamental question is: "Which characteristics of the configuration remain invariant during a continuous, gradual change and which do not?" In geometry teaching, work towards theorem-finding is to be based on answering this question through the continuous variation of geometrical configurations.

Discovering theorems by varying geometrical figures

4.2 EXAMPLES OF THEOREM FINDING

Continuous transformation processes are documented in this book by phase pictures or 'snapshots'; these can only imperfectly represent the dynamics. The examples are taken from lessons involving triangles and quadrilaterals. The choice is arbitrary; only some examples conform directly with the material content of current geometry syllabuses for lower secondary school. The constructions are generally not described and the accompanying theorems are generally not formulated.

Cabri-géomètre cannot handle all possible construction problems: for example to illustrate Morley's Theorem. This remarkable theorem, discovered by F Morley in about 1899 and proved in 1914, states that the three points of intersection of adjacent trisectors of the angles of *any* triangle form an *equilateral triangle*. Since this requires the *trisection* of a general angle it is not to achieveable with compasses-and-ruler and (so) not with Cabri-géomètre. Another kind of theorem which cannot be illustrated is one which involves invariance of area, since Cabri-géomètre's drag-mode cannot, in general, be constrained to hold an area invariant.

4.2.1 Examples from 'Triangle geometry'

Examples 1-3 refer to properties of special points of triangles. Particular special cases are included.

Example 1 *Orthocentre*
The altitudes of a triangle have a common point of intersection called the orthocentre, H (Fig. 4.1). Triangle ABC is constantly varied. The common point of intersection is preserved.

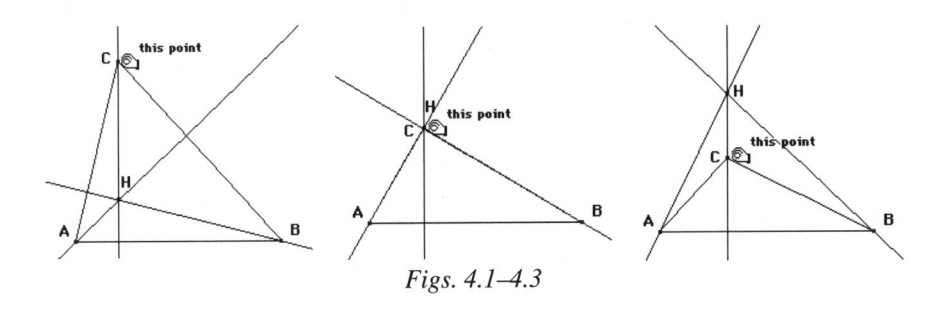

Figs. 4.1–4.3

Particular cases: the orthocentre, H,

- lies inside an acute-angled triangle (Fig. 4.1);
- coincides with the vertex of a right-angled triangle (Fig. 4.2);
- lies outside an obtuse-angled triangle (Fig. 4.3).

Example 2 *Circumcentre*

The perpendicular bisectors of the sides of a triangle have a common point of intersection called the circumcentre, M (Fig. 4.4).

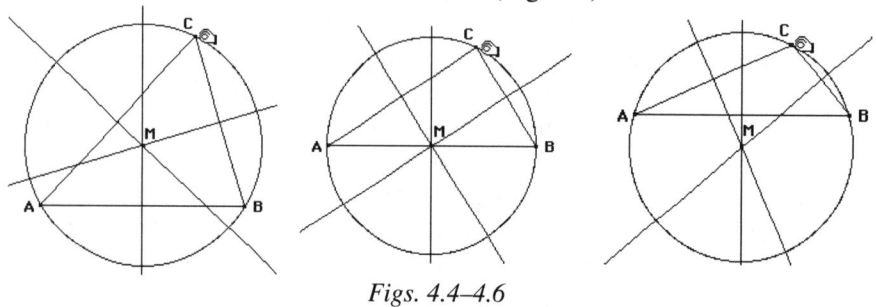

Figs. 4.4–4.6

The circumcentre, M,

- lies inside an acute-angled triangle (Fig. 4.4);
- coincides with the vertex of a right-angled triangle (Fig. 4.5);
- lies outside an obtuse-angled triangle (Fig. 4.6).

Example 3 *Medians and altitudes*

Figs. 4.7–4.9 show the following properties:
(1a) the common point of intersection of the medians (Fig. 4.7);
(1b) the common point of intersection of the altitudes (Fig. 4.7);
(2)   the partial concurrence of the medians and the altitudes of an isosceles triangle (Fig. 4.8);
(3)   the complete concurrence of the medians and the altitudes of an equilateral triangle (Fig. 4.9).

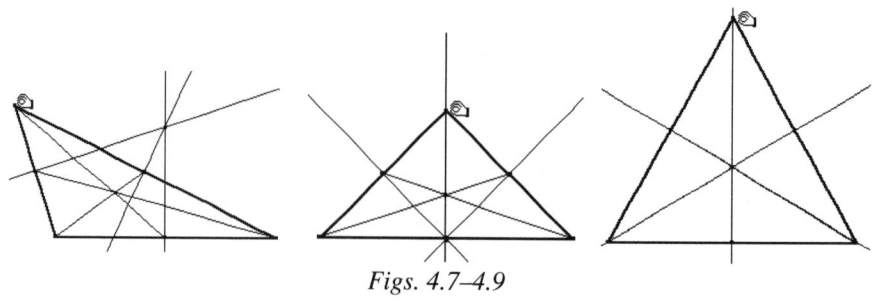

Figs. 4.7–4.9

Discovering theorems by varying geometrical figures

In Examples 4-6 below, a triangle is given, with three further line segments added which meet at a point and whose length measurements are displayed (for example, Fig. 4.10). Proceed as follows:

– Firstly, steadily vary the point and observe the length measurements, keeping the triangle fixed; this leads to a conjecture.
– Secondly, confirm the conjecture by steadily varying the given triangle (to test it in the general case).

Example 4 *Circumcentre*
The circumcentre is the point equidistant from the vertices (Figs. 4.10–4.15).

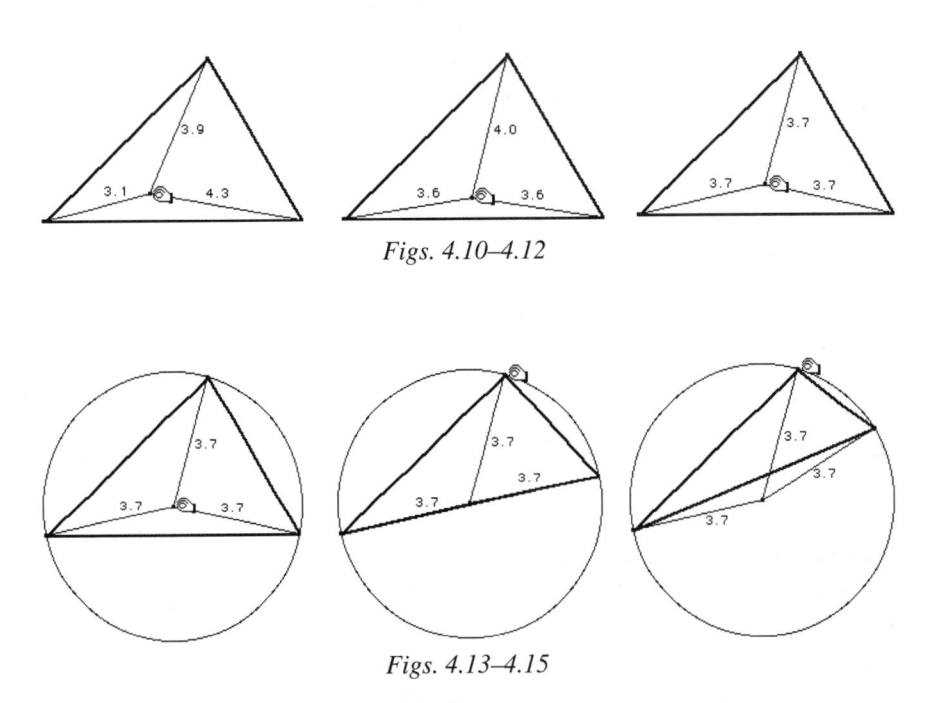

Figs. 4.10–4.12

Figs. 4.13–4.15

Example 5 *Viviani's theorem*
The sum of the distances to the three sides of an equilateral triangle from a point lying inside or on a side is constant and equal to the triangle's height (see Figs. 4.16–4.18). (This was introduced in Chapter 2, Example 4.)

77

Figs. 4.16–4.18

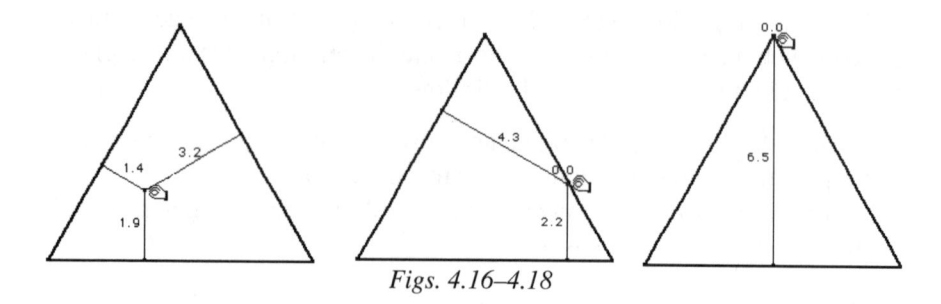

This theorem can be further generalised if the point is allowed to drift outside the triangle. In that case the perpendicular line which lies *totally* outside the triangle must be treated as having *negative* length. Figs. 4.19–4.21 show three such cases. The possible positions of the point are restricted by the fact that the feet of the perpendiculars must lie on the sides of the triangles, so the vertices provide boundaries (Figs. 4.20–4.21). The geometric locus of all limiting positions forms an interesting figure (for the reader to investigate).

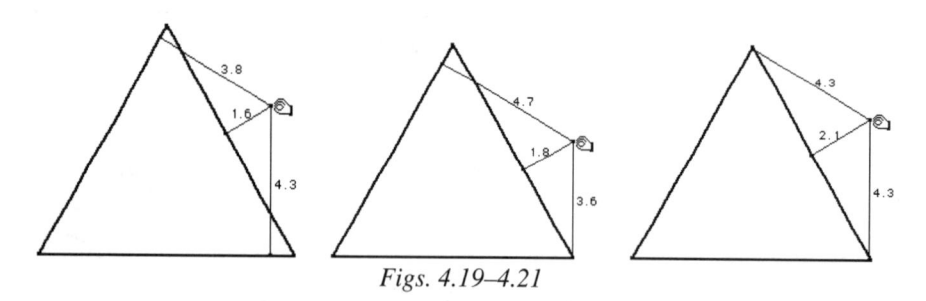

Figs. 4.19–4.21

Example 6 *Van Schooten's theorem*
If a point moves along the circumcircle of an equilateral triangle, the sum of the distances from the point to its two "neighbouring" vertices is always equal to the distance to the third vertex (Figs. 4.22–4.23). Diminishing or increasing the size of the equilateral triangle makes no difference – it scales the lengths proportionately (Figs. 4.24–4.25).

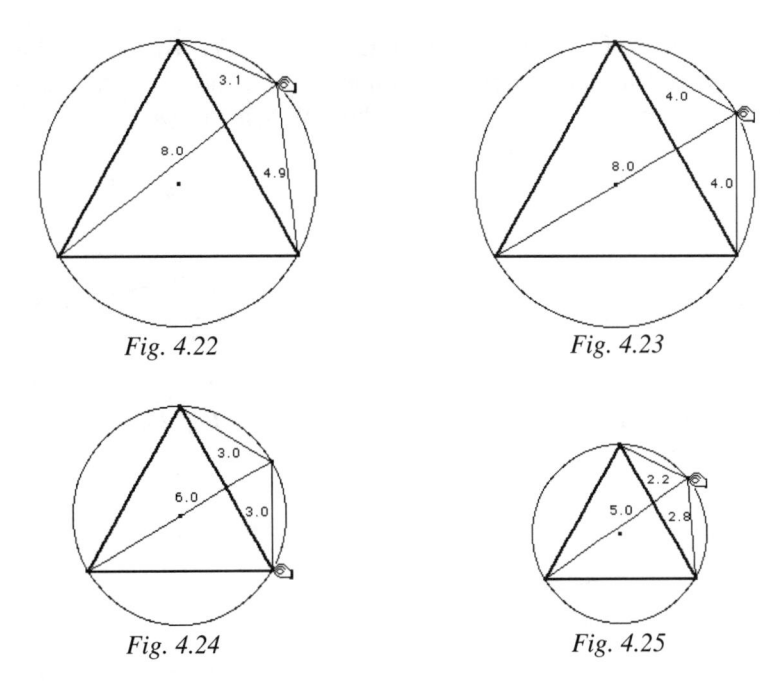

Discovering theorems by varying geometrical figures

Fig. 4.22

Fig. 4.23

Fig. 4.24

Fig. 4.25

Example 7 *Minimal characteristic of the pedal triangle [a harder example]*

We draw an acute-angled triangle ABC and a general triangle whose vertices U, V, W lie on its sides (Fig. 4.26). U, V, W can be moved along the sides of the triangle ABC.

When does the perimeter of the inner triangle UVW have a minimum? This is known as Fagnano's problem.

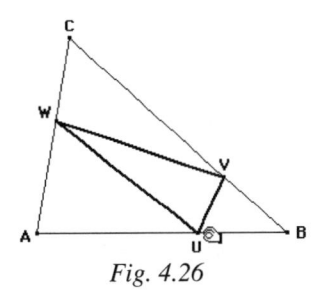

Fig. 4.26

We reflect WU in AC and reflect VU in BC. The length $[U_1WVU_2]$ which equals the perimeter of the inner triangle is not minimal for the given posi-

79

tion of U because U_1WVU_2 is not straight (Fig. 4.27). To achieve this, we shift V and W onto the straight line connecting U_1 and U_2. Figs. 4.27–4.28 show V being moved; Figs, 4.29–4.30 show W being moved.

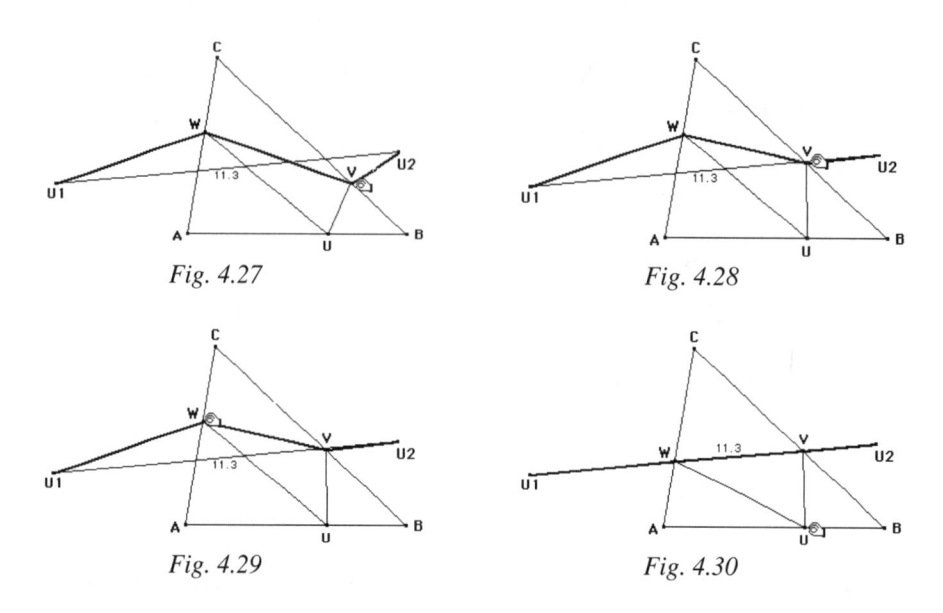

Fig. 4.27

Fig. 4.28

Fig. 4.29

Fig. 4.30

Then we shift U and check whether the length $[U_1U_2]$ can be decreased, and repeat the minimisation of triangle UVW (Fig. 4.31) until a smallest value for $[U_1U_2]$ is reached experimentally (Fig. 4.32).

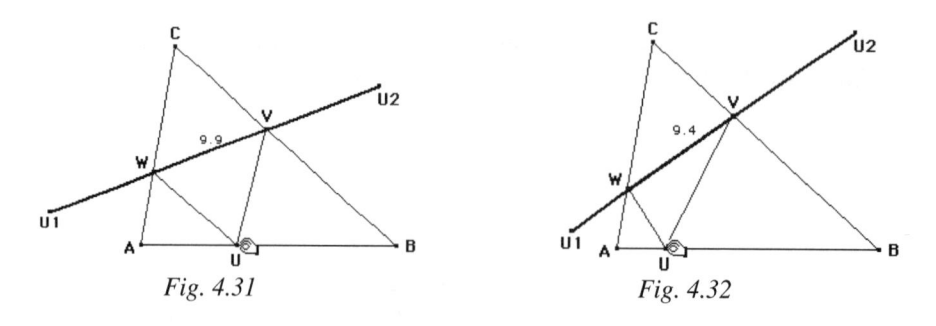

Fig. 4.31

Fig. 4.32

It is found that the minimal position is achieved when U, V, W are the feet of the altitudes of triangle ABC. This is when UVW is the pedal triangle of ABC. For this particular figure the minimum length is 9.4 units. We now *construct* the pedal triangle and continuously change triangle ABC and observe that its perimeter (represented in the figures as a line segment) remains minimal – signified by the line segments remaining in a straight line (Figs. 4.33–4.34).

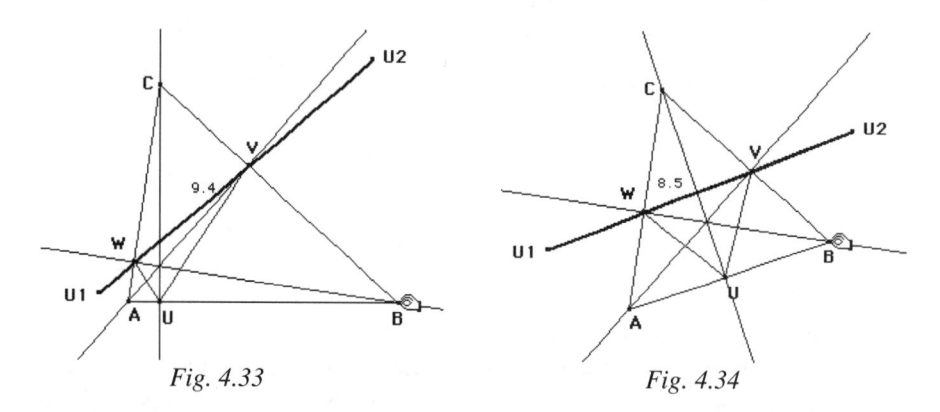

Fig. 4.33 *Fig. 4.34*

In Examples 8-9 below, configurations based on equilateral triangles are changed to more general ones through continuous transition.

Example 8 *Fermat point*
The configuration consists of an equilateral triangle as base with an equilateral triangle drawn outwards on each of its three sides. The vertices of the outer triangles are each connected to the opposite vertex of the base triangle (Fig. 4.35). The connecting lines intersect each other at one point. Where they intersect they create six equal angles of 60°. We vary the base triangle so that it is no longer equilateral (Fig. 4.36). The property that the lines have a common point of intersection remains invariant, provided that the outer three triangles remain equilateral. Likewise, the property that the angles of intersection are equal (60°) is invariant (Figs. 4.37–4.38). What about the lengths of the connecting lines?

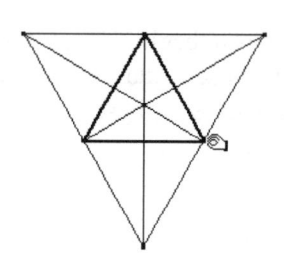

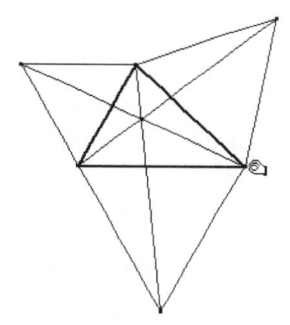

Fig. 4.35–4.36

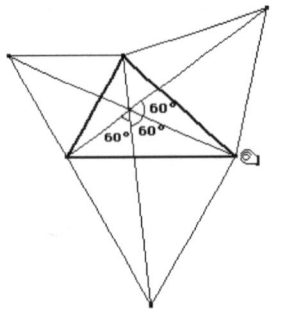

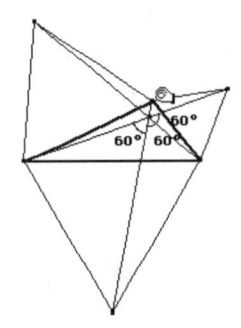

Fig. 4.37–4.38

Example 9 *Napoleon's theorem*
We begin with an equilateral triangle. On each side an equilateral triangle is constructed, pointing outwards (as in Example 8). The centroids of these three outer triangles are joined to form the 'mid-triangle' (Fig. 4.39). In our initial configuration the mid-triangle is equilateral. When we change the base triangle to any acute-angled or obtuse-angled triangle (Figs. 4.40–4.41), the shape of the mid-triangle does not change – it is always equilateral. (Is this still true if the three equilateral triangles are constructed pointing *inwards*?)

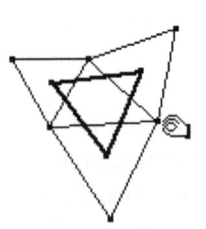

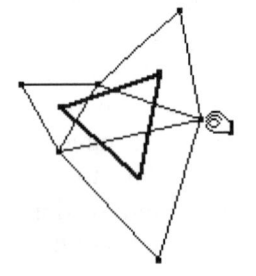

Fig. 4.39–4.41

Note: a more common but equivalent approach is to begin with an equilateral triangle and construct on the middle third of each side a one-third size equilateral triangle pointing out, then join the apexes of these three smaller triangles. This is left to the reader to investigate.

4.2.2 Examples from 'Quadrilateral geometry'

In analogy to Example 9 above, we can erect four squares on the sides of a quadrilateral (Fig. 4.42) and consider the quadrilateral formed by joining the centres of the four erected squares (which we call the 'mid-quadrilateral'), and its diagonals (Fig. 4.43).

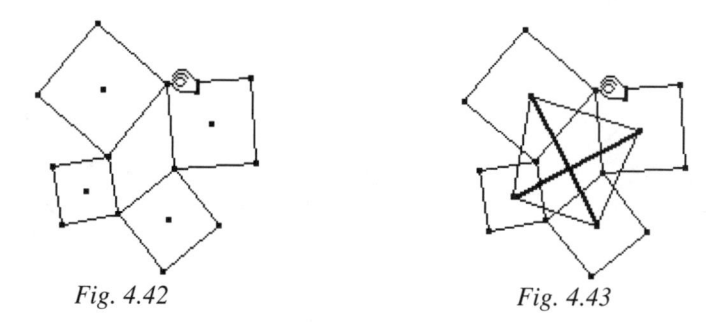

Fig. 4.42

Fig. 4.43

Example 10 *Diagonals of the mid-quadrilateral*
We take Fig. 4.42 again and construct the mid-quadrilateral as before (Fig. 4.43) and hide the sides of the mid-quadrilateral (Fig. 4.44). Then we change the base quadrilateral continuously and find that the diagonals of the mid-quadrilateral remain equal in length and orthogonal (Fig. 4.45), establishing Aubel's theorem.

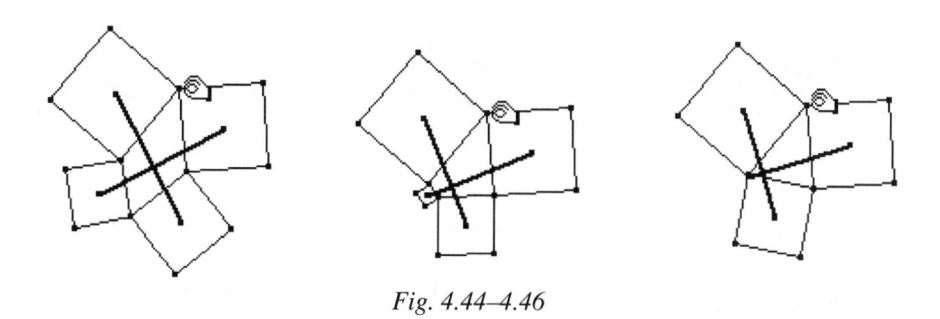

Fig. 4.44–4.46

Note: Quadrilaterals with orthogonal diagonals are characterized by the fact that the two sums of the squares of opposite sides are equal. Mid-quadrilaterals are a particular kind of such quadrilateral, because the diagonals have equal lengths. Quadrilaterals can only be mid-quadrilaterals if their diagonals have both properties (i.e. equality of length and orthogonality).

In Fig. 4.46 we move on to the limiting case of the base quadrilateral being deformed to a triangle, which gives us the corresponding statement for a triangle with squares erected on its sides.

Instead of starting from a general quadrilateral base, we could begin with a square base (Fig. 4.47) and generalise from the square to a general quadrilateral base (Figs. 4.48–4.49).

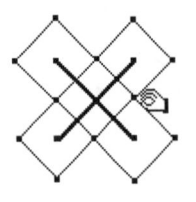

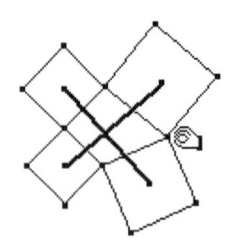

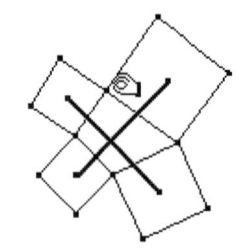

Fig. 4.47–4.49

Example 11 *Angle-halving quadrilateral*
A quadrilateral is formed inside a parallelogram by bisecting the angles of the parallelogram (Fig. 4.50); this inner quadrilateral is right-angled. The parallelogram is lengthened until two opposite corners of the inner rectangle lie on the opposite sides of the parallelogram (Fig. 4.51).

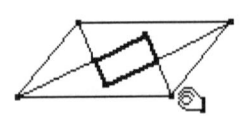

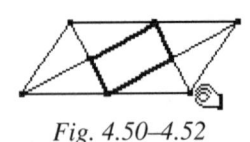

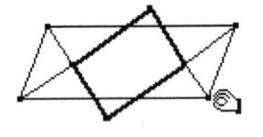

Fig. 4.50–4.52

What condition is satisfied by the sides of the parallelogram?

The lengthening of one side of the parallelogram is continued further (Fig. 4.52).

Also, whatever variation of the parallelogram's sides and angles is undertaken, the rectangular form of the quadrilateral remains invariant (not illustrated).

We specialise the parallelogram to a rhombus, then to a rectangle and finally to a square, and observe the respective change of the angle-halving rectangle (Figs. 4.53–4.55).

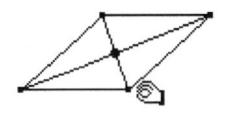

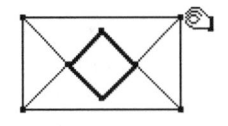

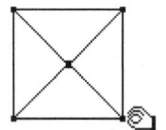

Discovering theorems by varying geometrical figures

Figs. 4.53–4.55

Distortion of the parallelogram to an isosceles trapezium reveals a kite-quadrilateral with two right-angled opposite vertices (Fig. 4.56).

What condition must be satisfied by the vertices of the isosceles trapezium so that the kite-quadrilateral becomes a single point? (Fig. 4.57).

When the vertices are moved beyond the single point position an (inverted) kite-quadrilateral reappears (Fig. 4.58).

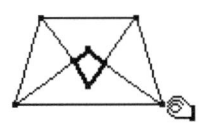

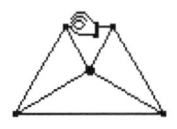

Figs. 4.56–4.58

Example 12 Pythagoras' Theorem
Continuous deformation of a figure illustrates Pythagoras' Theorem (Figs. 4.59–4.62).

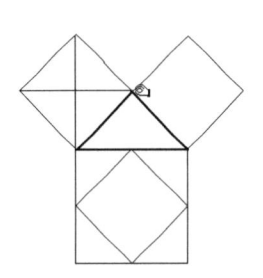

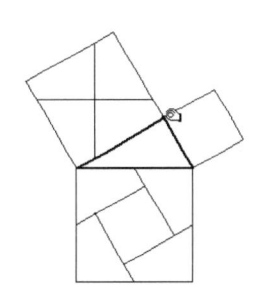

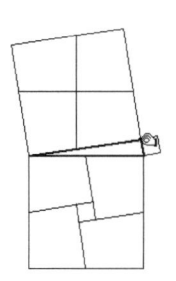

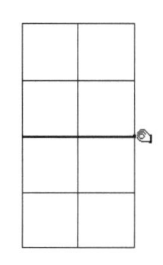

Figs. 4.59–4.62

4.3 MOTIVATION FOR PROVING THEOREMS

Eradicating doubt is no longer the main argument for proving theorems since moving diagrams are very convincing! One might try to motivate stu-

85

dents to undertake formal proofs by saying that the screen has only a finite number of pixel points and so only a finite number of figures can be represented, but this will not succeed or stimulate; the students are totally convinced that what they have seen is universally true. The educator must therefore emphasise another aspect of proving, namely, to answer the question *"Why* does this property hold?" This forces the students to find arguments from their geometrical and intuitive knowledge. The students' line of argument is elevated to a logical level, beyond the screen figure level, which is a principle aim of mathematics education. This convergent thinking can counterbalance the growth in divergent thinking implicit in the exploratory work seeking invariance properties of figures.

CHAPTER 5

Angle theorems as invariance properties

> *As lines, so loves oblique, may well*
> *themselves in every angle greet,*
> *But ours so truly parallel,*
> *though infinite can never meet.*
> Andrew Marvell

5.1 INTRODUCTION

The concepts angle and right-angle, and the many angle theorems based on them, form an important part of the geometry curriculum for lower secondary school (the didactical difficulties of which will not be discussed here). Teaching angle concepts has received attention from educators and some technical assistance in the form of software has been developed but the topic remains problematical. There are many different ways of thinking about angles – directed (amount of turn), undirected, convex, non-convex, pair of intersecting lines, area bounded by two line segments. The students do not seem impressed by the educators' efforts to unravel these possible approaches; experience indicates that students' problems are much more basic and lie in measuring angles and in handling a protractor. This weakness has negative consequences for the learning processes of discovering angle theorems inductively and solving construction tasks involving angles. There are negative consequences in adult life too for angle measurement by craftsmen, handymen and leisure time navigators, etc. In the following, we concentrate our thoughts on inductive theorem–finding in school geometry by graphic exploration. Such a graphics system must have the following three facilities:

(a) Automatic measurement of lines, angles and polygonal areas.
(b) Drag–mode which allows the user to continuously change a previously constructed configuration (with measurement values automatically updated).
(c) A file system to allow preparation of a library of changeable configurations.

Continuous variation is a real–time process, somewhat like having a user-controlled cartoon. A graphics system incorporating real-time variation, such as Cabri-géomètre, can promote the discovery of elementary geometrical theorems by assisting the user to answer the fundamental question: "Which properties of the configuration remain invariant during continuous variation?" (the Invariance Principle). We assume, mainly from our knowledge of cognitive psychology and motivation theory, that with this method of theorem–finding, the discovery and learning of geometrical facts will become more attractive and more effective for the student. The following draft didactic scheme results:

1. Construction of an appropriate configuration by the student or loading of a library configuration supplied by the teacher.
2. Variation of the configuration (generation of different cases, limiting cases, specialisation and generalisation; variation of position and size). Addressing the questions: What changes? What does not change? What are the conjectures or hypotheses?
3. Strengthening of conjectures through further variation and measurement where appropriate (e.g. observing variation in angles, with their measurements).
4. Formulation of statements (as universal statements or theorems).

The above scheme allows evidence to be gathered in support or in refutation of a conjecture. The question then arises: How should this conjecture be proved? Figure-specific proofs at a logical level could be added now (often done at secondary level).

For example:
- In Fig. 5.1 we see that the perpendicular bisectors of two sides of a triangle meet (we do not think about proving this as it is quite evident).
- Then we add the third perpendicular bisector and see that it too is concurrent (this we *do* try to prove) as it is unexpected and not obvious (Fig. 5.2).

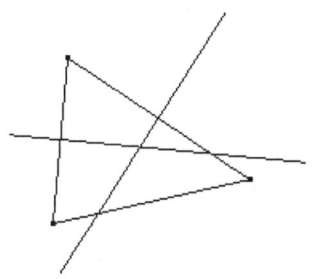

Fig. 5.1 Two perpendicular bisectors meeting

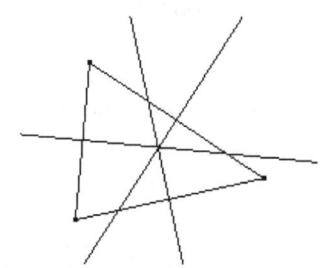

Fig. 5.2 Three perpendicular bisectors concurrent

The usual formal deductive proofs of theorems may then be proceeded to, where feasible and desirable. The motivation for formal proof is seldom present since students are already fully convinced. It is better to ask *why* the inductively found proposition is true and *how* it relates to other known facts. This can place the new statement in the hierarchy of known statements.

In Section 5.2, elementary angle theorems found in school syllabuses are developed as invariance statements according to the above scheme.

Some means of sequencing learning steps are:

- *isolating* (e.g. separately studying perimeter of a rectangle and then area of a rectangle); or *relating* (e.g. from the beginning studying both perimeter and area of a rectangle together, and asking how the two are related);
- *specialising* (e.g. going from polygon to rectangle to square) or *generalising* (e.g. going from square to rhombus to parallelogram);
- *concretising* or
   abstracting.

The linear learning sequences in Section 5.2 represent a proposal for a teaching experiment using a computer as the interactive medium.

5.2 Learning–sequences based on angle theorems as invariance statements

Cabri-géomètre allows us to construct an unorientated angle as a measurable object by giving three points (in the sequence: side point, vertex, side point). It is not possible to define an angle in terms of a non–ordered pair of semi-infinite lines, since semi-infinite lines are not permitted as elementary objects in current versions of Cabri-géomètre. Reflex angles cannot be constructed either – their production as non–orientated angles would require marking of a further point in the interior of the angle, in order to differentiate between the reflex angle and the corresponding acute angle. This limitation is a particular disadvantage in describing rotations with reflex angles, and in formulating the angle-at-the-centre theorem for the case where the angle at the circumference is obtuse (Fig. 5.3), and in the concept of interior angle for non–convex polygons.

Restricting angle measurement to whole numbers complies with the accuracy of protractors used in schools and with the angle values normally given in school construction tasks. This accuracy level occasionally leads to inconsistency when the rounding-off of calculated values to whole numbers leads to incorrect statements. For example, for the angle-at-the-centre theorem if the centre angle has a measure which is an odd integer then the measure of the angle at the circumference cannot be displayed as exactly half that at the centre (Fig. 5.4).

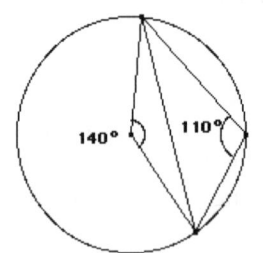

Fig. 5.3 Cabri-géomètre reporting wrong angle

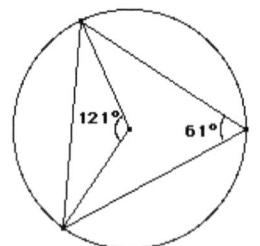

Fig. 5.4 Cabri-géomètre round-off error

On the one hand, such problems can be eased by the use of decimal places; on the other hand, such inconsistencies may motivate students to look for more formal arguments in order to resolve them.

In the following, the continuous variation of the configurations is documented by phase–pictures ('snapshots'). The basic point to be moved is marked by the 'gripping hand' icon. The statements of invariance which result are, in general, not explicitly formulated in our examples – they are self-evident or should be familiar to the reader. Studying on-screen measurements of angles is a good way to discover invariances.

Angle theorems as invariance properties

5.3 LEARNING SEQUENCE 1

– Angle
– Angle measurement
– Angle addition

5.3.1 Angle

We start from the right–angle (Fig. 5.5) and reduce this to an acute angle (Fig. 5.6) or increase it to an obtuse angle (Fig. 5.7). In the limiting positions we get a 180° angle or straight line (Fig. 5.8) or a zero–angle (Fig. 5.9) where the angle's two sides lie on top of each other. One can vary the positions of the two side points and the vertex, and thereby alter the size or position of the angle or the lengths of the angle sides.

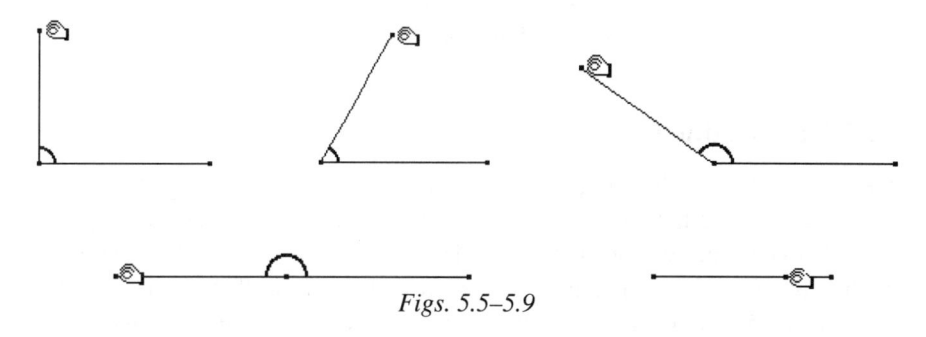

Figs. 5.5–5.9

5.3.2 Angle measurement

We now measure the angle on-screen. As we open up the zero–angle to the 180° angle, increasing angle values are shown (Fig. 5.10–5.14); there are, of course, just 181 possible integer angle values which can be reported.

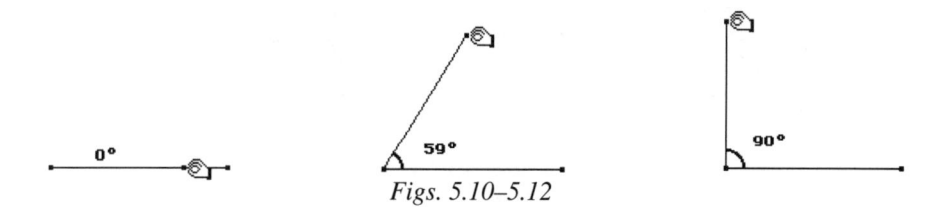

Figs. 5.10–5.12

Discovering Geometry with a Computer

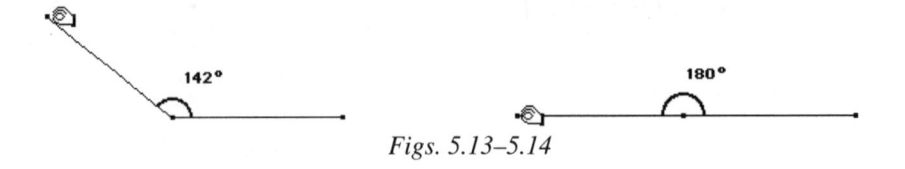

142°   180°

Figs. 5.13–5.14

We can simulate a protractor by using a side point constrained to move on a semi–circle (Fig. 5.15).

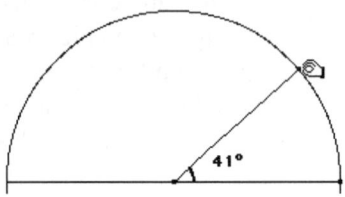

41°

Fig. 5.15 Screen–protractor

5.3.3 Angle addition

We construct an angle using two lines which meet at a vertex. Then we construct an intermediate line from the vertex to some other arbitrary point lying between the two original lines. This forms two new angles. The sum of the values of the two smaller angles is seen to equal that of the original angle (Fig. 5.16). This is confirmed by variation of the intermediate line (Fig. 5.17), and, furthermore, by variation of the position of the two original lines and their intersection point (not illustrated).

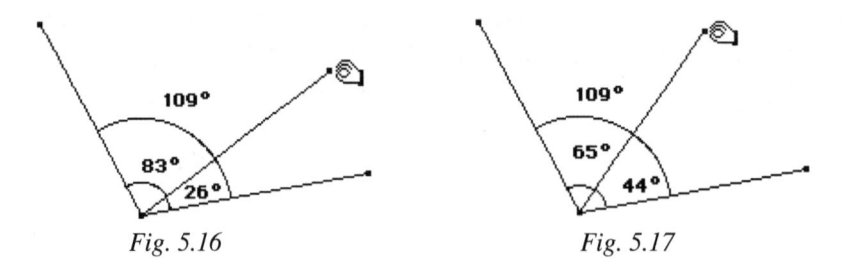

109°   109°
83°   65°
26°   44°

Fig. 5.16   *Fig. 5.17*

5.4 LEARNING SEQUENCE 2

– Adjacent angle
– Vertical angle

Angle theorems as invariance properties

- Angles on skew and parallel lines
- Corresponding angles on parallel lines
- Alternate angles on parallel lines
- Opposite angles on parallel lines
- Angles in a parallelogram.

5.4.1 Adjacent and vertical angles

We start with two lines forming a T shape with two adjacent right-angles at the common point (Fig. 5.18). One line is rotated about the common point until the zero–angle position is reached (Figs. 5.19–5.20). From the previous work on angle addition we see that the sum of the two adjacent angles equals the straight line angle. This can then be checked by measuring (not illustrated).

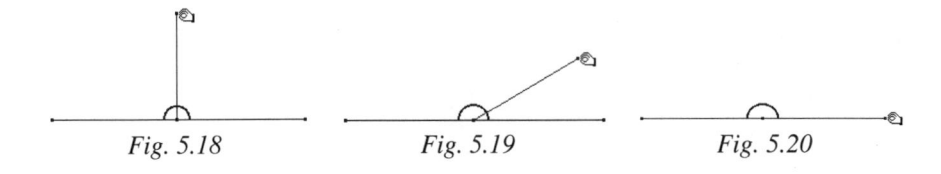

Fig. 5.18   *Fig. 5.19*   *Fig. 5.20*

In Figs. 5.21–5.22, starting from two straight lines at right-angles, adjacent and vertically opposite angles are seen to be related. Adjacent angles are complementary i.e. add to 180°; vertically opposite angles are equal. The relationships are maintained during variation of the figure. These properties can be studied without or with measurements displayed.

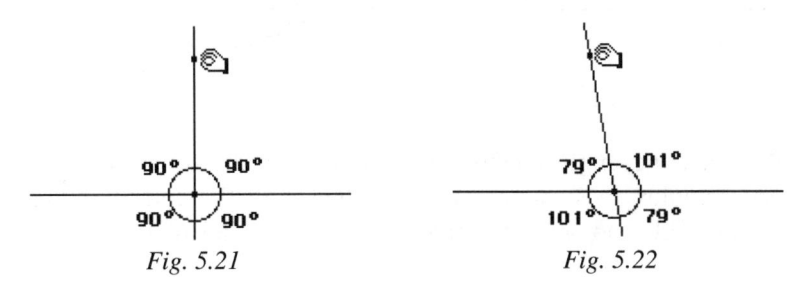

Fig. 5.21   *Fig. 5.22*

5.4.2 Angles on skew and parallel lines

Two given non-parallel straight lines are intersected by a third line (the transversal); we mark and measure all angles at the two intersection points

93

(Fig. 5.23). There are many different pairings of angles possible (adjacent, opposite, corresponding, ... using one or the other or both points of intersection). We change the position of one of the given lines until it is parallel to the second line (Figs. 5.24–5.25) and note that now the corresponding adjacent and vertically opposite angles match at the two points of intersection ... which leads us to noting that corresponding angles are equal, alternate angles are equal, opposite angles are complementary. Now the position of the transversal is changed. The relationship properties of the relevant angles are invariant (Fig. 5.26).

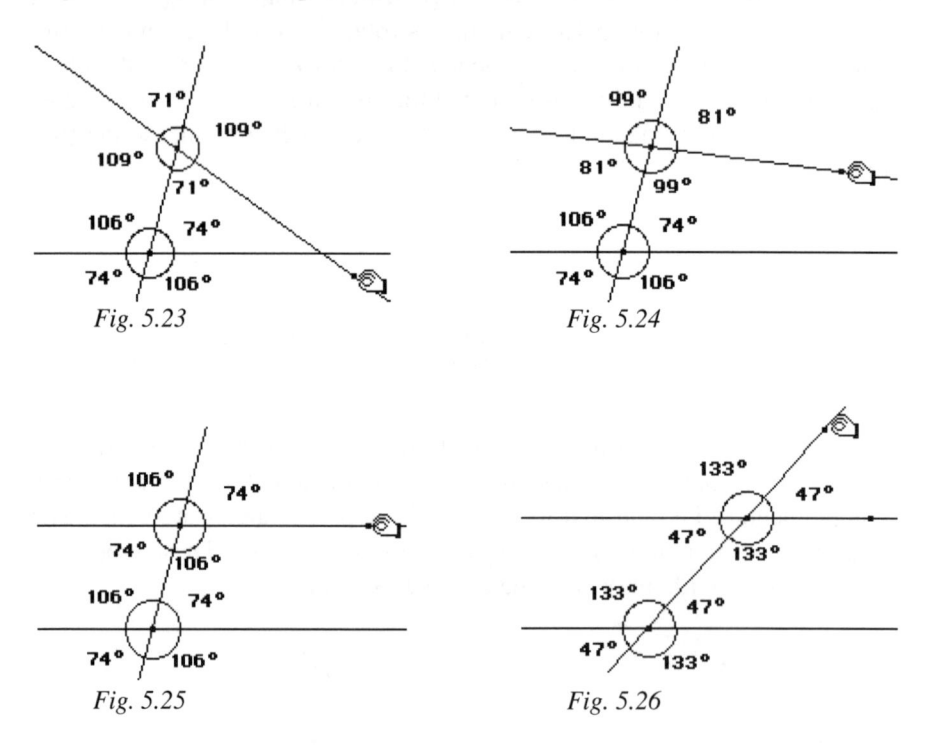

Fig. 5.23

Fig. 5.24

Fig. 5.25

Fig. 5.26

In the following, the angle configurations discussed above, which may appear very complex to students, are simplified with the corresponding angle, alternate angle and opposite angle aspects treated in isolation.

5.4.3 Corresponding, alternate and opposite angles on parallel lines

We draw a corresponding angle configuration as in Fig. 5.27 (without and with measurements). This is continuously varied (Figs. 5.28–5.30); the relationship between the corresponding angles is preserved.

Angle theorems as invariance properties

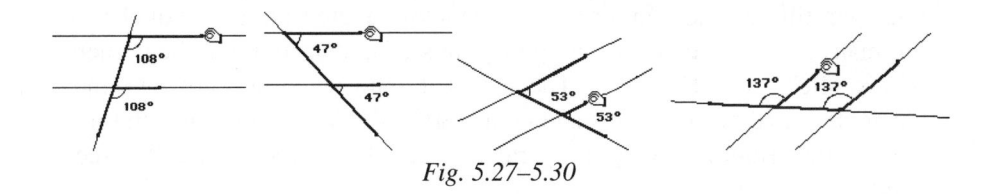

Fig. 5.27–5.30

The corresponding angle theorem, the adjacent angle theorem, and the properties of angle measurement, together form a deductive basis for many angle statements. For this reason, a further investigation is given here, involving continuous parallel shifting: one of the parallel lines is shifted until the corresponding angles coincide (Figs. 5.31–5.33).

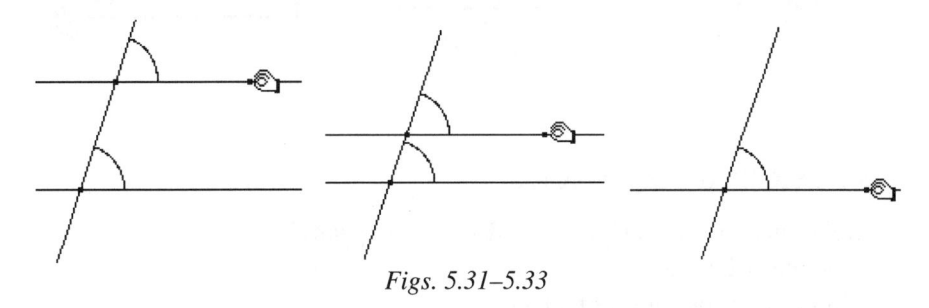

Figs. 5.31–5.33

The other angle relationships produced by two parallel lines and a transversal, can be treated in the same way.

5.4.4 Angles in a parallelogram

Two parallel lines are intersected by two other parallel lines and the four angles at the four points of intersection are marked (Fig. 5.34). We recognise equal angles and measure them if necessary to confirm this (Fig. 5.35). Here we can set up motivating tasks requiring students to fill in missing information: "Which angles are the same?", "Complete the missing angle values.", etc. (Fig. 5.36).

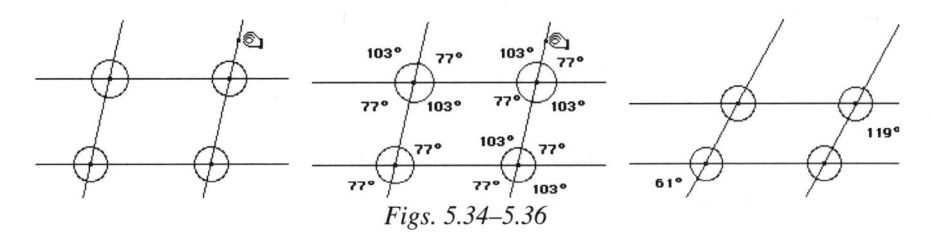

Figs. 5.34–5.36

Such gap-filling tasks can form an introductory exercise or a way of checking many angle statements. The four points of intersection are the corners of a parallelogram (Fig. 5.37). The angles at opposite corners are the same; the angles in adjacent corners add up to 180°. We hide the lines which form the parallelogram and vary the parallelogram (Fig. 5.38) until it is a rectangle (Fig. 5.39).

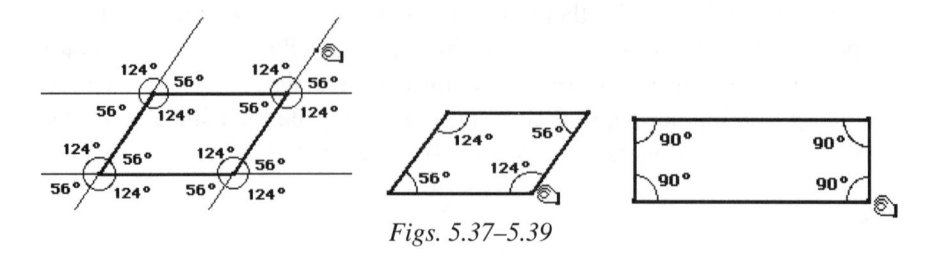

Figs. 5.37–5.39

5.5 LEARNING SEQUENCE 3

- Relationship between angles and sides of a triangle
- Isosceles triangle
- Sum of interior angles of a triangle
- Sum of interior angles of a convex quadrilateral
- Sum of interior angles of a polygon
- Exterior angle theorem of a triangle
- Exterior angle sum of a convex polygon

5.5.1 Relationship between angles and sides of a triangle

We draw a triangle with its internal angles marked, and measure all angles and sides (Fig. 5.40). We vary the triangle (Figs. 5.41–5.42).

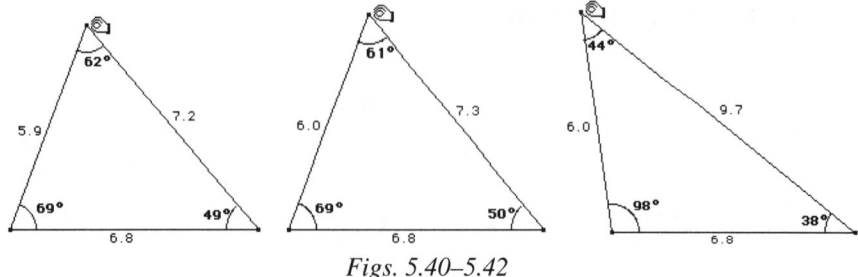

Figs. 5.40–5.42

Angle theorems as invariance properties

The following properties are invariant:

- The longest (middle, shortest) side is opposite the largest (middle, smallest) angle, and vice versa.
- If one angle is obtuse, the other two are always acute.

We vary the figure so that two angles (or sides) are the same size and see that then two corresponding sides (or angles) are also the same size (Fig. 5.43). We can even make all three sides the same length; then each angle is 60° and all the sides are equal (Fig. 5.44).

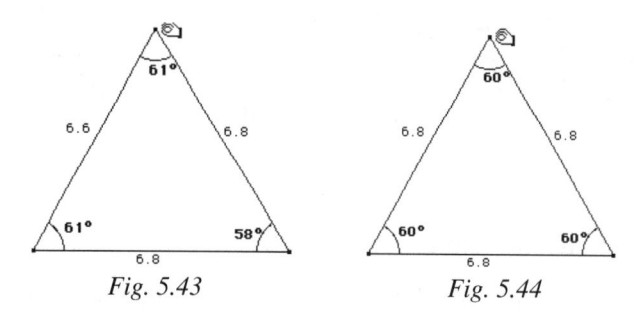

Fig. 5.43 *Fig. 5.44*

5.5.2 Isosceles triangle

We make two sides of a triangle the same length; the other we call the base. The two base angles are equal. We draw in the perpendicular bisector of the base and see that it passes through the third vertex (Fig. 5.45). The position of the vertex of the isosceles triangle which is linked to the perpendicular bisector is now varied (Figs. 5.46–5.49); the two base angles always remain equal. Amongst other things, we notice that an increase of the angle at the vertex being moved corresponds to a reduction in the base angles (half coming from each), and similarly a reduction in the angle at the vertex corresponds to an evenly shared increase in the base angles. The sum of the three angles remains the same, namely 180°, which can be seen particularly clearly in the special case of the equilateral triangle (Fig. 5.47). Thus we have found the interior angle sum theorem for the isosceles triangle.

Figs. 5.45–5.49

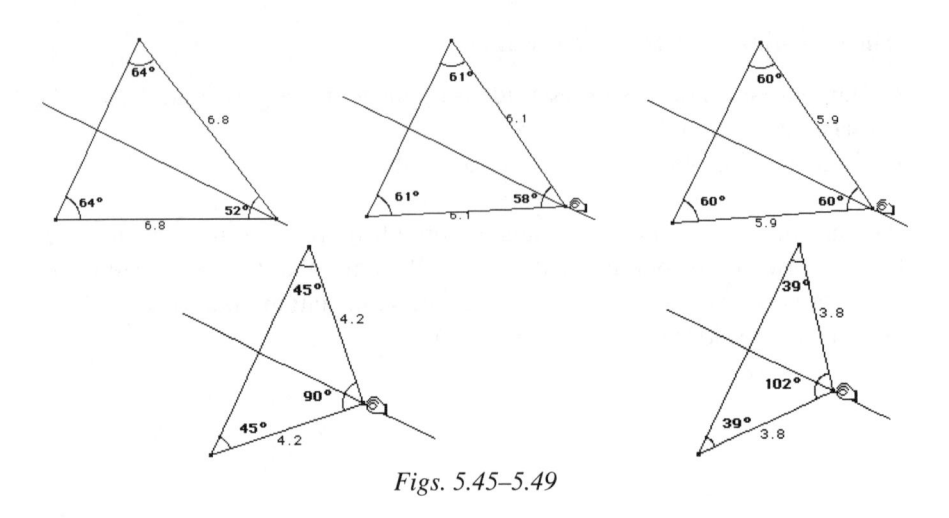

5.5.3 Sum of interior angles of a triangle

We start from a square which is divided into two congruent parts by a diagonal (Fig. 5.50). After hiding one part, an isosceles right–angled triangle remains (Fig. 5.51), which is then changed into a non–isosceles right-angled triangle (Fig. 5.52). The changed base angles complement one another, increasing and decreasing in balance, to sum to 90°. The right–angled triangle is now made into a non–right–angled triangle (Fig. 5.53). The sum of the three angles remains 180°.

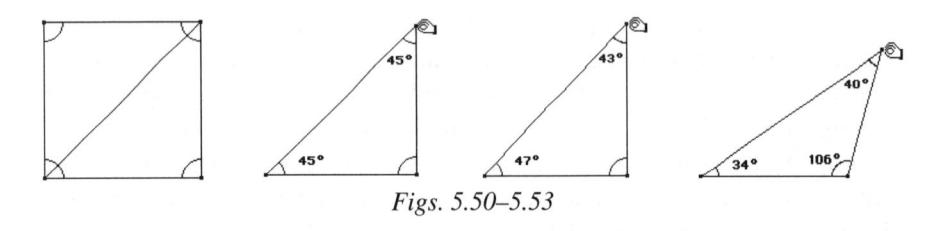

Figs. 5.50–5.53

This whole process is reversible: we can produce an isosceles right-angled triangle by transforming any given triangle.

In order to provide an explanation for the angle sum being 180° using known statements, we form a total angle out of the three internal angles, utilising the corresponding angle property for parallel lines; this gives a straight line (Figs. 5.54–5.55).

Figs. 5.54–5.55

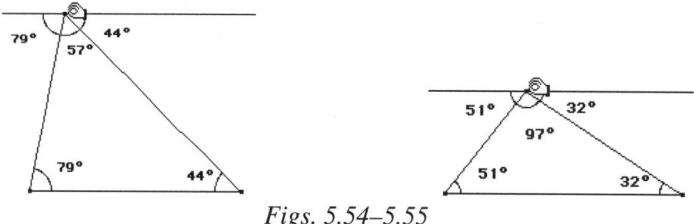

5.5.4 Sum of interior angles of a convex quadrilateral

We start from a square (Fig. 5.56) and generalise it step by step so that the angles at just two neighbouring vertices are changed at each step. We recognise that the increase in size of one angle corresponds to a decrease in size of the other (Figs. 5.57–5.59); the sum of the interior angles stays equal to four right angles (360°). In Fig. 5.60 we make one corner flatten to a straight line (180°) and get a triangle as a limiting case.

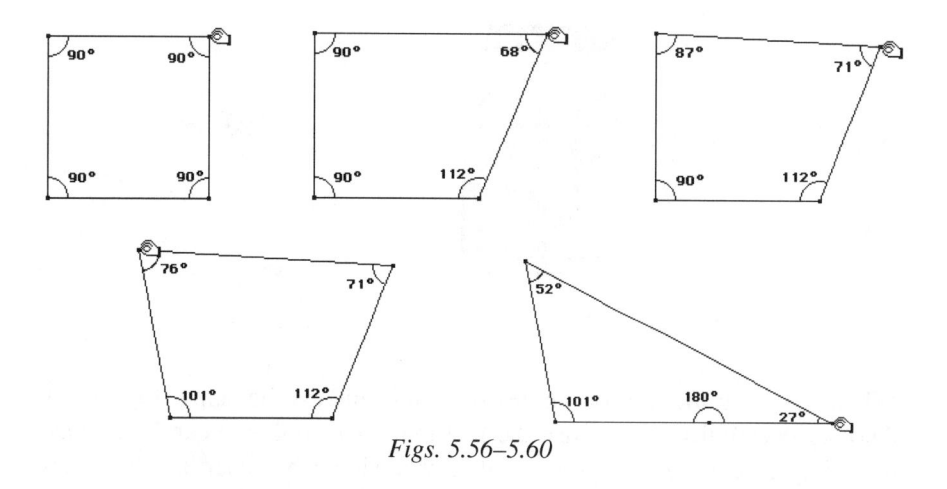

Figs. 5.56–5.60

Another way of perceiving the sum of the interior angles of a quadrilateral lies in dividing it into two triangles (Fig. 5.61). We gradually change a convex quadrilateral so that a triangle appears as an intermediate position (Fig. 5.62) and then continue to form a non–convex quadrilateral (Fig. 5.63). The sum of the four angles remains 360° throughout. The concept of angle is quietly broadened here to the reflex, which cannot be constructed with the current version of Cabri-géomètre.

Discovering Geometry with a Computer

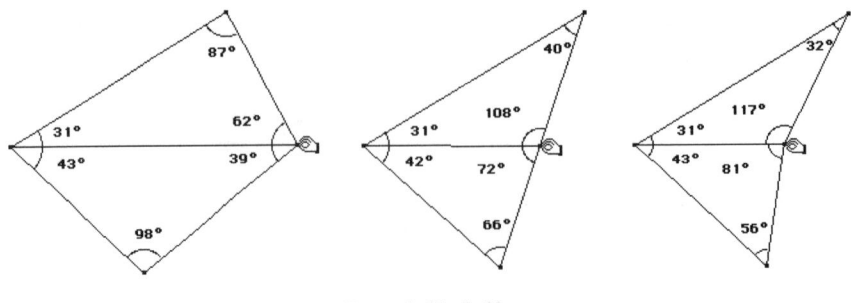

Figs. 5.61–5.63

5.5.5 Sum of interior angles of a polygon

We add up the interior angles of a convex polygon and get $(n-2) \times 180°$ (Fig. 5.64 with $n=7$, where an on–screen calculator can be used to help with the addition; Fig. 5.65 with $n=5$)

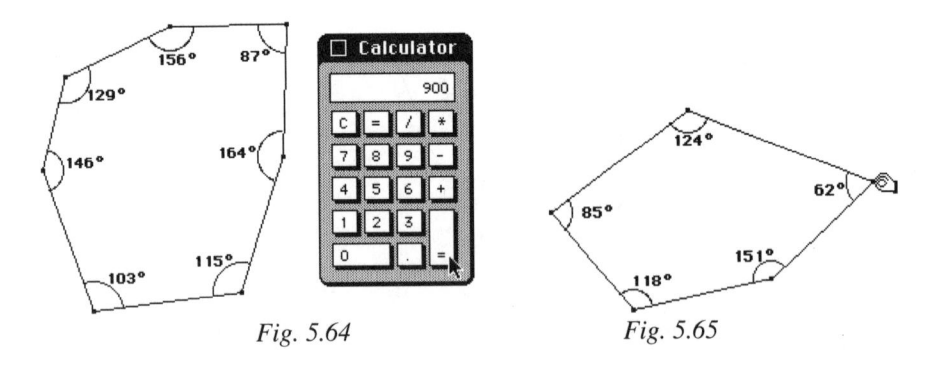

Fig. 5.64   *Fig. 5.65*

When changing the interior angles the total remains the same (Fig. 5.66). Another possibility is to utilise the fact that every polygon can be divided into simpler convex polygons (with fewer sides) by inserting diagonals (Fig. 5.67). The transition to a non–convex polygon can follow; once again the size of the sum of the interior angles remains the same (Fig. 5.68).

Angle theorems as invariance properties

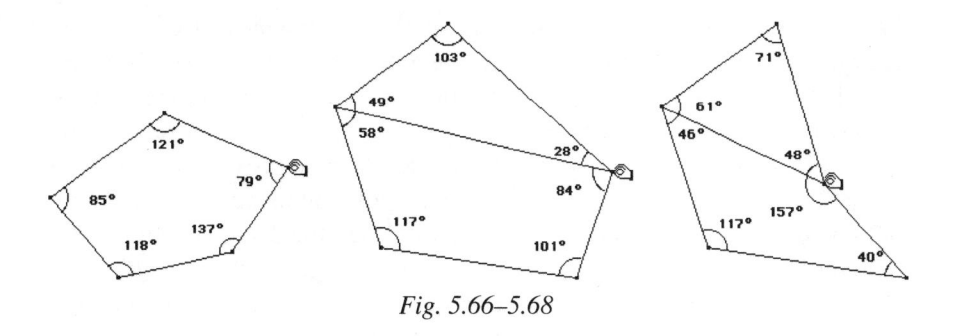

Fig. 5.66–5.68

5.5.6 Exterior angle theorem of a triangle

In Figs. 5.69–5.70, a triangle with an exterior angle marked is varied. The sum of the two interior opposite angles remains equal to that of the exterior angle.

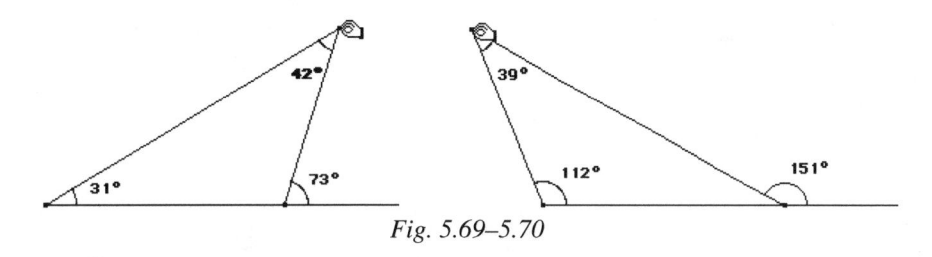

Fig. 5.69–5.70

5.5.7 Exterior angle sum of a convex polygon

We begin with a triangle ($n=3$) and measure its exterior angles; their sum is 360° (Fig. 5.71). The usual angle balancing as the triangle is deformed leads to the invariance of the sum of the exterior angles (Fig. 5.72).

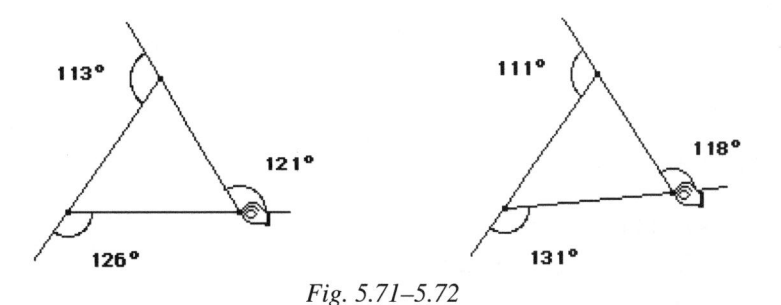

Fig. 5.71–5.72

We now proceed to consider the sum of the exterior angles of a convex polygon: the sum of the exterior angles of a polygon always gives 360° (Fig 5.73 for *n*=5). Reduction of one exterior angle causes a matching increase of one (or two) neighbouring exterior angles, so that the sum remains the same (Fig. 5.74). The following consideration aims to prove it (by complete induction): we cut a triangle from a convex polygon using a diagonal (Fig. 5.75, with *n*=5). The sum of the exterior angles at the diagonal endpoints is thereby increased by the value of the exterior angle at the cut off vertex (Fig. 5.76). The sum of the exterior angles of the n-sided polygon and that of the (*n*–1)–sided polygon, are therefore equal.

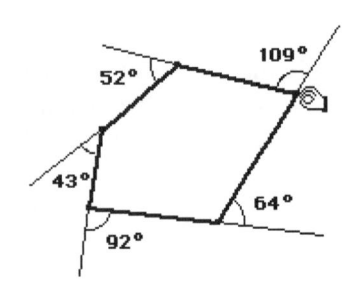

Fig. 5.73–5.74

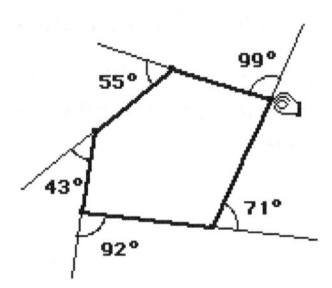

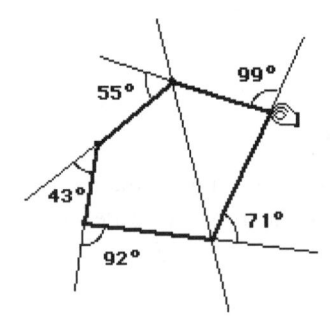

Fig. 5.75–5.76

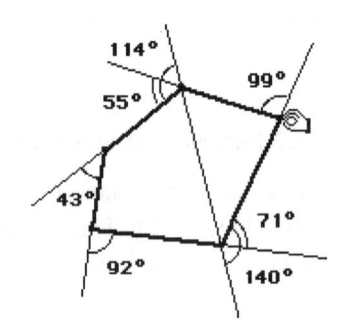

5.6 LEARNING SEQUENCE 4

- Angles in an isosceles triangle
- Angles in a cyclic quadrilateral
- Thales' theorem
- Theorem of angle at the circumference

Angle theorems as invariance properties

- Converse of theorem of angle at the circumference
- Chord–tangent angles
- Normal angles

5.6.1 Angles in an isosceles triangle

See Section 5.5.2.

5.6.2 Angles in a cyclic quadrilateral

We start with a rectangle and its circumcircle (Fig. 5.77), generalise the rectangle to a quadrilateral with the same circumcircle and observe that the angles in the opposite corners always add up to 180° (Figs. 5.78–5.80). This is a property of the cyclic quadrilateral. Also, we notice that the value of the angle at the point being moved round the circle remains constant. This is a property of a chord to a circle.

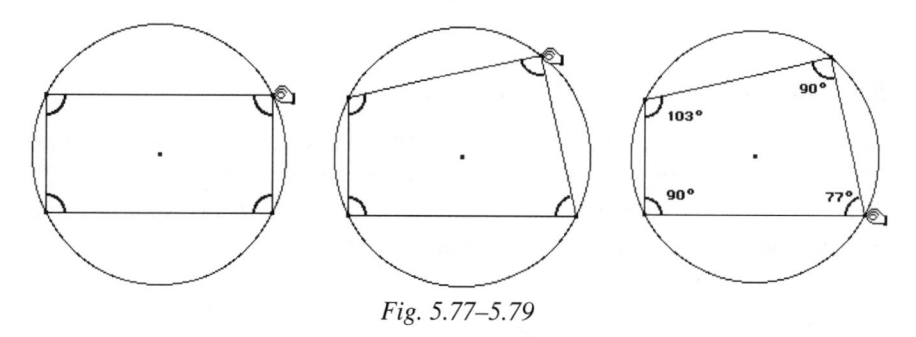

Fig. 5.77–5.79

We could, of course, start with a general quadrilateral (rather than a square) as in Fig 5.81, whose variation shows the invariance of the sums of opposite pairs of angles (Fig. 5.82).

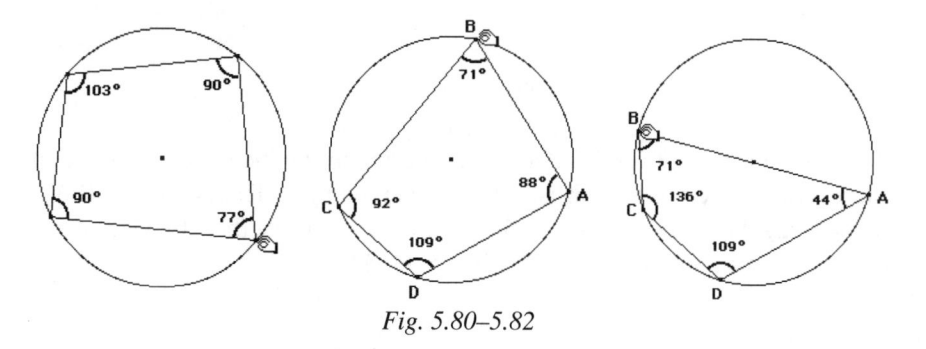

Fig. 5.80–5.82

Discovering Geometry with a Computer

5.6.3 Thales' theorem

A triangle is inscribed in a circle. The triangle has the diameter as one of its sides. The vertex opposite the diagonal is moved along the circle (Fig. 5.83). We notice that the size of the angle at this vertex does not change (Fig. 5.84). This can be studied without and with the angle measurements displayed. Fig 5.85 serves as a vivid demonstration of this result.

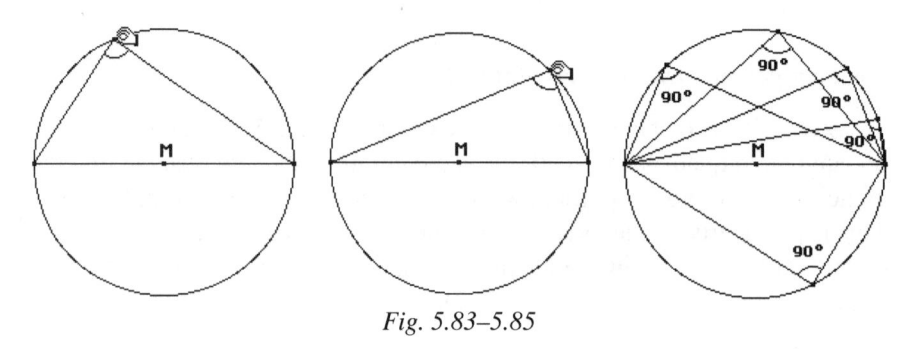

Fig. 5.83–5.85

5.6.4 Theorem of angle at the circumference

Here we generalise a figure. We replace the diameter in the Thales' figure (Fig. 5.86) by a chord which is *not* a diameter by dragging B (Fig. 5.87). Then we keep moving the vertex opposite the chord (C) (Fig. 5.88); the measurement of the angle at C does not change.

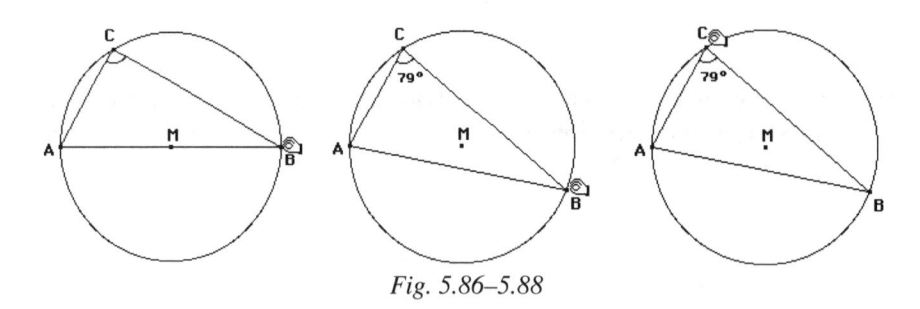

Fig. 5.86–5.88

The relationship between the angle at the circumference and the angle at the centre is likewise established by generalising a Thales' figure (Figs. 5.89–5.92). With fixed angle at the centre, we test whether the size of the circumference angle is invariant (Figs. 5.93–5.94).

Figs. 5.89–5.91

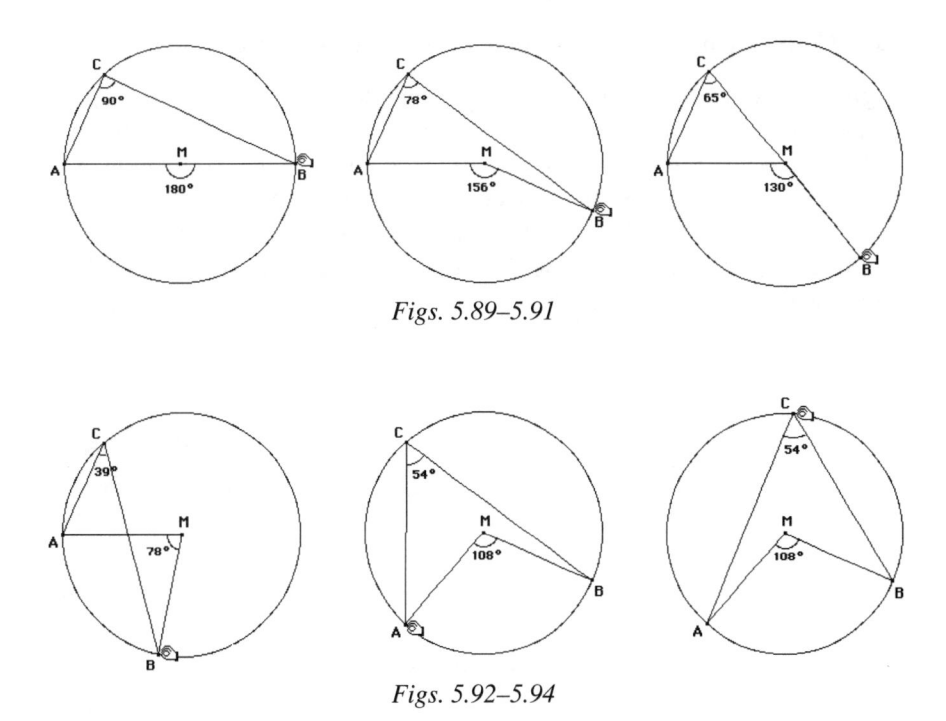

Figs. 5.92–5.94

In Fig. 5.95 chord AB is of a fixed length (being a copy of the vertical line segment to the left of the circle). When the chord's position is varied the corresponding angle at the circumference remains the same (Fig. 5.96). For every chord (AB) there are two angles at the circumference (Fig. 5.97), which sum to 180°. With this result concerning the angles at the circumference at vertices C and C', we obtain the theorem of cyclic quadrilaterals (opposite angles are complementary).

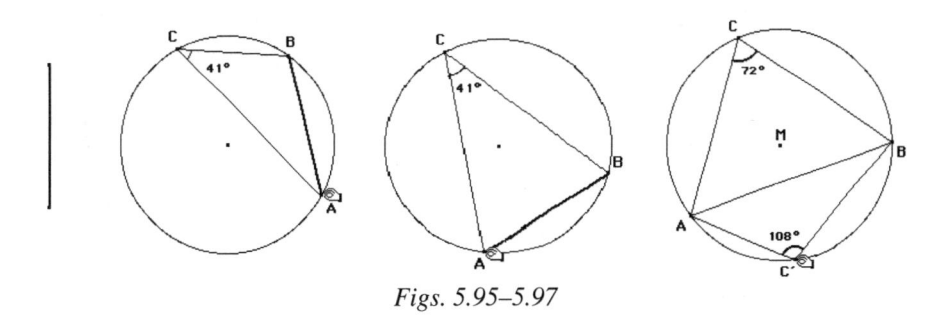

Figs. 5.95–5.97

105

5.6.5 Converse of theorem of angle at the circumference

A given fixed angle is constructed. On a separate free line defined by point A and one other (un-named) point, a point C is placed. A free point B is chosen such that CA and CB define an angle equal to the fixed angle (Fig. 5.98). We move the free line containing CA. The interactively produced locus of C shows that C runs along an arc of a circle for which AB is a chord (Fig. 5.99). (Hint: choose the Locus command from the **Miscellaneous** menu, click on C and then rotate the free line by dragging the un-named point slowly.)

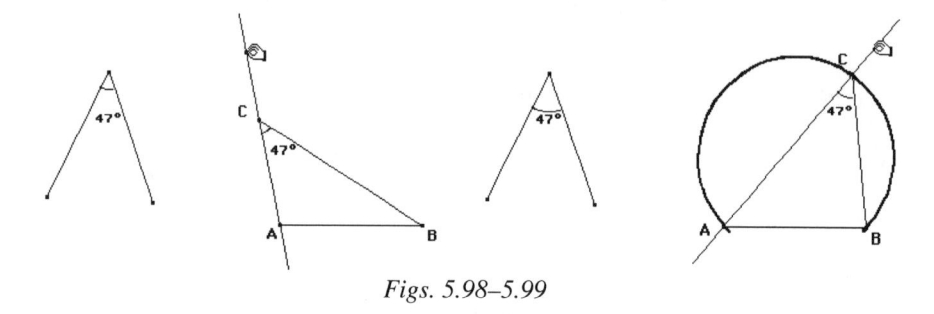

Figs. 5.98–5.99

5.6.6 Chord-tangent angles

A chord (Fig. 5.100: chord CB) changes into a tangent in the limiting position (Figs. 5.101–5.102) and the angle at the circumference becomes the chord-tangent angle. During variation of the chord, the angle(s) at the circumference and chord-tangent angle(s) remain equal (Figs. 5.103–5.104).

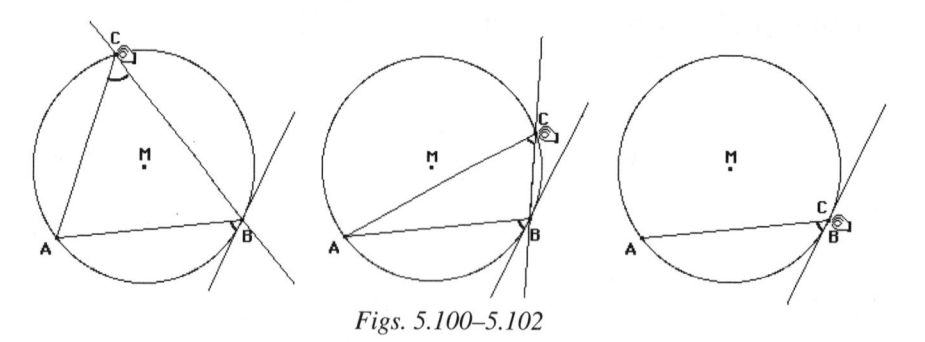

Figs. 5.100–5.102

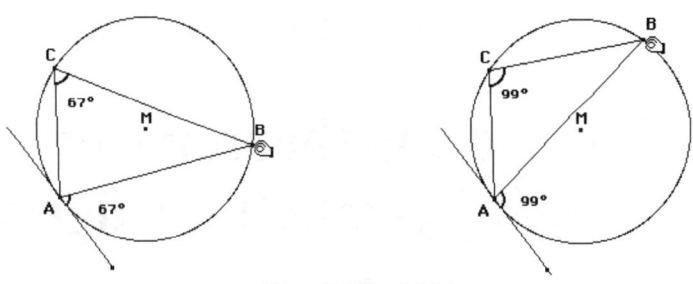

Figs. 5.103–5.104

5.6.7 Normal angles

Because we can use the normal angle theorem as a basis for the chord-tangent angle theorem, we will include it here. We construct two angles as in Fig 5.105, whose defining sides stand in pairs at right-angles (and so are normals to each other). Variation shows that the normal angles are either the same size or add up to 180° (Figs. 5.106–5.107).

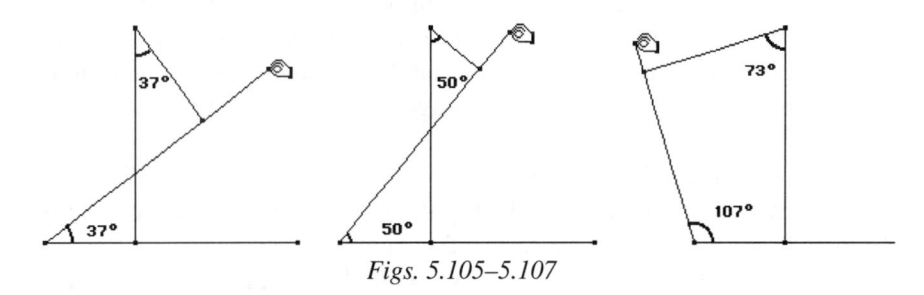

Figs. 5.105–5.107

A special case of the normal angle theorem appears in the rotation of a straight line about a centre. The angle of rotation is always equal to the angle between the straight line in its new position and in its initial position (Figs. 5.108–5.109).

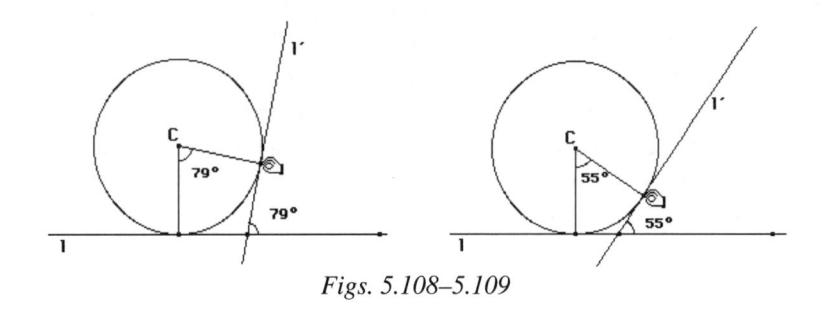

Figs. 5.108–5.109

CHAPTER 6

Discovering theorems by generalising geometrical figures

All generalisations are dangerous.
Alexandre Dumas fils

6.1 INTRODUCTION

An essential form of learning in school mathematics is the study of how to solve problems. An important component of this is the study of heuristic methods. Owing to its iconic and thus easily accessible presentation, synthetic elementary geometry is traditionally acknowledged as a training ground for heuristic thinking (even though practice may not reflect this!).

One heuristic method is **generalising**, which occurs in two essentially different forms when applied to geometry theorem-finding. One is abstracting from individual cases in the form of geometric drawings to an abstract formulation (called theorem finding by **incomplete induction** e.g. concluding that **any** triangle's perpendicular bisectors are concurrent, from studying many examples). The other is deriving a new statement (or theorem) by transition of one specific situation into a more general situation (called theorem finding by **conceptual generalisation** e.g. concluding from the square's properties that **any** parallelogram's diagonals bisect each other). The latter mode of gaining cognition can be put into concrete form at the iconic level as is illustrated in Diagram 6.1: in order to proceed from specific theory to general theory there is a transition to a specific graphic configuration, representing one instance of the geometric statement, to a more general graphic configuration, and finally the general theory is abstracted. (The abstraction processes intervene naturally, if statements have to be read and verbally expressed from individual graphic constructions.)

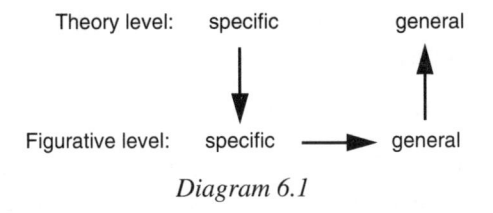

Discovering theorems by generalisering geometrical figures

Theory level: specific → general

Figurative level: specific → general

Diagram 6.1

Example 1

We begin with a rectangle and mark the midpoints of the sides. A line is drawn from each vertex to an opposite midpoint so that these four lines form a parallelogram in the centre of the rectangle (Fig. 6.1). We generalise the shape of the initial rectangle to a parallelogram and observe the effect (Fig. 6.2), establishing the generality of Varignon's theorem.

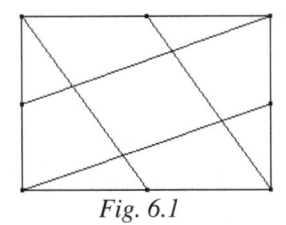

Fig. 6.1 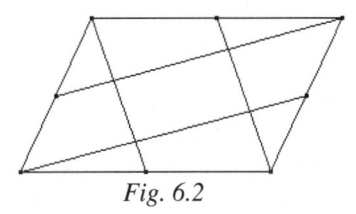   *Fig. 6.2*

Conceptual generalisation is especially productive if the theorem to be generalised can be represented by a symmetric configuration, and if the valdity of the theorem can be immediately seen from this configuration, based on its symmetrical characteristics. The question then arises about the invariance of properties observed in the symmetric configuration. Will the property hold if the configuration is transformed to a less symmetric or asymmetric (and thus more general) form? (The Principle of Invariance.) These invariant characteristics then form the more generalised statement. In this way the learner acquires from what is known something that was previously unknown – a realisation of the general didactic principle: 'From the known to the unknown'.

The following four stage heuristic process results from these considerations:

(1) Produce a symmetric configuration which represents an obvious geometric statement.

(2) Destroy symmetrical characteristics of this configuration by transformation.

Discovering Geometry with a Computer

(3) Check whether the original statement, or part of it, is still valid for the less symmetric or asymmetric configuration.

(4) Formulate the generalised statement.

The motivation for applying heuristic methods such as *incomplete induction* and *conceptual generalisation*, as well as their efficiency, depend to a large extent on the type of graphical construction tools available. The traditional tools for such construction (ruler and compasses ...) have considerable deficiencies but these can be remedied by using interactive computer graphic systems. Examples of this follow in the next section.

A useful facility is the option for redefining a constructed object as a basic object. Such an object can then be referred to and is (completely or partially) freely movable. We illustrate this with an example concerning the third (dependent) vertex of an equilateral triangle which is determined by the other two (independent) vertices.

Example 2
We start with the two given points and need to find the third by geometric construction. Using software we simulate the usual equilateral triangle construction with compasses utilising the intersection of two circles to locate the third vertex (Fig. 6.3). Then this third vertex is redefined as a basic object by pointing to it and selecting the appropriate **Miscellaneous** menu option **Redefine an object** (Fig. 6.4); it is now 'tangible' (Fig. 6.5) and can be freely moved by the user (Fig. 6.6). In this way, the equilateral triangle is transformed into 'any' triangle. Without this redefinition, although the two defining vertices at the base of the triangle could be freely moved this third vertex could not. It would always move in response to the others' movements – to preserve the equilateral nature of the triangle. A statement, based on the equilateral triangle (Fig. 6.7: perpendicular bisectors are concurrent), can be generalised to a statement about triangles in general (Fig. 6.8) using the above redefinition and drag-mode transformations.

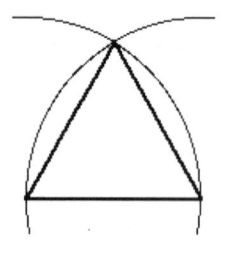

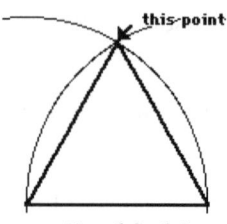

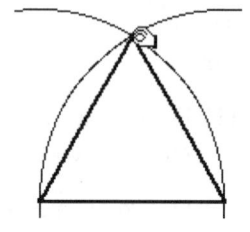

Fig. 6.3–6.5

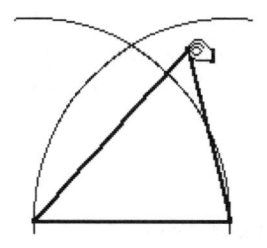

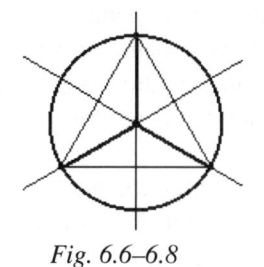

Discovering theorems by generalisering geometrical figures

Fig. 6.6–6.8

6.2 EXAMPLES OF GENERALISING THEOREM FINDING

We illustrate the generalising of theorem finding by using examples from the theory of the *n*-gon for $n = 3, 4, 6$. The selection is somewhat arbitrary. For reasons of space the dynamic generalising of the symmetric configuration is generally documented by just two phase pictures: (i) the symmetric initial configuration and (ii) a picture of the generalised configuration (with the 'hand' grabbing the basic point which is moved). The resulting generalised statements are not explicitly formulated as they will be self-evident or already known to the reader or, if not, will provide a stimulus for follow-up work.

By providing symmetric configurations as files which can be retrieved from the computer's memory by the user, the need for action (1) in the four stage heuristic method described in Section 6.1 becomes unnecessary. This has the advantage that theorem finding is not hindered by the lack of drawing competence.

6.2.1 Examples from triangle theory

Example 3 *Perpendicular bisectors and circumcircle*
Based on the symmetry properties of the equilateral triangle, the perpendicular bisectors of the sides intersect each other at one point which is the same distance from each vertex; the vertices are thus located on a circle, the circumcircle of the triangle, the centre of which is the intersection of the perpendicular bisectors (Fig. 6.7). The third vertex is converted to a basic point (as discussed above) and dragged off from its symmetric position (Fig. 6.8); the concurrency of the perpendicular bisectors is invariant. The other lines which traverse the triangle (i.e. angle bisectors, medians, altitu-

des), which coincide with the perpendicular bisectors in the case of the equilateral triangle, also have a common intersection point which is equidistant from the vertices, but the latter property is *not* invariant under asymmetrisation. However, other characteristics *are* invariant ...

Example 4 *Triangulation through the medians*
The equilateral triangle is cut into six congruent triangles (therefore all of equal area) by its medians (Fig. 6.9). When the symmetry is destroyed, it is obvious that two of the six triangles must remain equal in area to each other due to their having equal bases and the same altitude (Fig. 6.10). Further consideration confirms the fact that all six smaller triangles are equal in area.

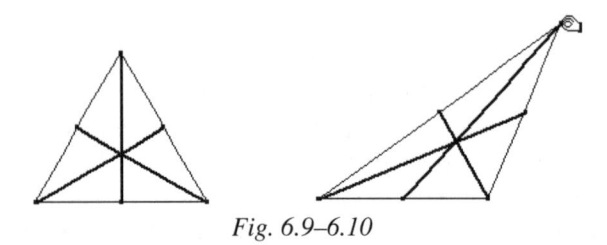

Fig. 6.9–6.10

Example 5 *Inscribed triangle with minimum perimeter – Fagnano's problem*
Starting with an equilateral triangle, we draw the pedal triangle – the triangle formed by the three feet of the altitudes. We reflect two sides of the pedal triangle about the sides of the initial triangle. This produces a representation of the perimeter of the pedal triangle as three joined line segments (Fig. 6.11). (This same problem was approached in a different way in Chapter 4, Example 7.)

It is clear that the perimeter will be minimal if the line segments lie in one straight line. A larger perimeter can be obtained when a foot is generalised to a free movable point on one side of the triangle (Fig. 6.12). The minimal property of the pedal triangle is preserved if we change the equilateral triangle into 'any' scalene triangle (Fig. 6.13). If, however, starting with an equilateral triangle, we draw the triangle joining the **midpoints** of the three sides, we find that it does *not* remain minimal with regard to the perimeter when the equilateral triangle is distorted (Figs. 6.14–6.15).

Figs. 6.11–6.13

Figs. 6.14–6.15

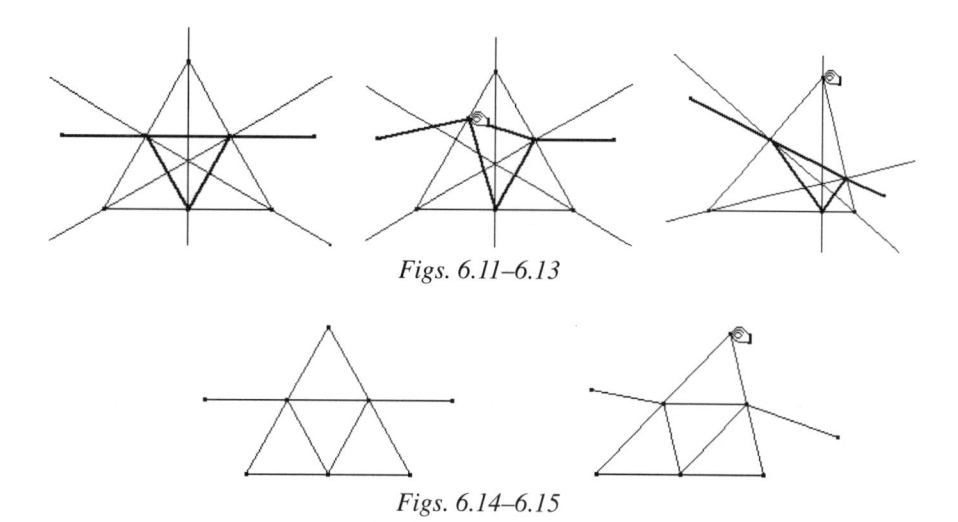

Example 6 *Fermat point*
On the sides of a central equilateral triangle three equilateral triangles are drawn facing outwards. The circumcircles of these three outer triangles intersect each other at one point (Fig. 6.16). Each side of the central triangle subtends an angle of 120° at this intersection point. This property is preserved if we transform to *any* central triangle provided that each of its internal angles is less than 120° (Fig. 6.17). (A variant of this was met in Chapter 4, Example 8.)

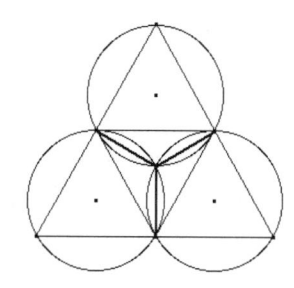

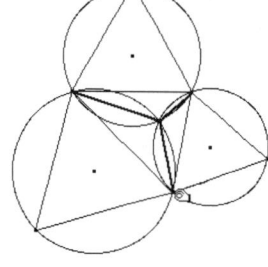

Figs. 6.16–6.17

Example 7 *Inscribed square*
We wish to draw a square inscribed in an equilateral triangle. First we draw a square which has only *two* vertices on the sides of the equilateral triangle, in symmetric position (Fig. 6.18). (This uses the solution strategy of omitting some conditions).

Figs. 6.18–6.20

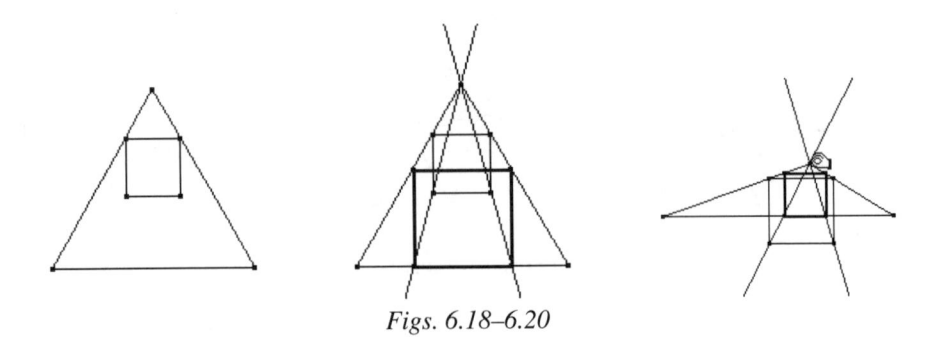

An enlargement of this square, with centre the appropriate vertex of the triangle, produces the inscribed square (Fig. 6.19). This solution can be generalised (Fig. 6.20).

6.2.2 Examples from quadrilateral theory

Example 8 *Mid-side quadrilateral*
A 'mid-side quadrilateral' is drawn in a (base) square by joining the midpoints of the base square's four sides. This too is a square, having half the area of the base square itself (Fig. 6.21). If we transform the base square into a parallelogram, the mid-side quadrilateral changes from a square into a parallelogram – also of half the area (Fig. 6.22). If the symmetry of the base parallelogram is destroyed to yield a general quadrilateral, the parallelogram shape of the mid-side quadrilateral is nevertheless preserved (Fig. 6.23), a result known as Varignon's theorem, and which we introduced in Example 1.

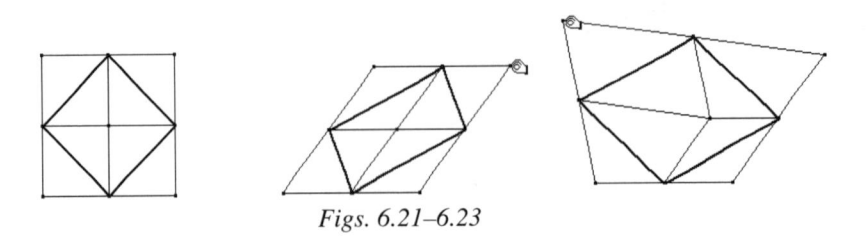

Figs. 6.21–6.23

Example 9 *Internal angle sum of a quadrilateral*
Beginning with a base square, a triple point reflection produces a tiling figure (Fig. 6.24) with 360° as its internal angle sum. The base square has its four internal angles marked (90° each of course) and these same angles

are replicated in the centre of the figure showing that the four add to a full turn (360°). If we generalise the base square to a right-angled trapezium the sum of its internal angle is invariant (Fig. 6.25). Furthermore, this remains so if the trapezium is changed to any convex quadrilateral (Fig. 6.26).

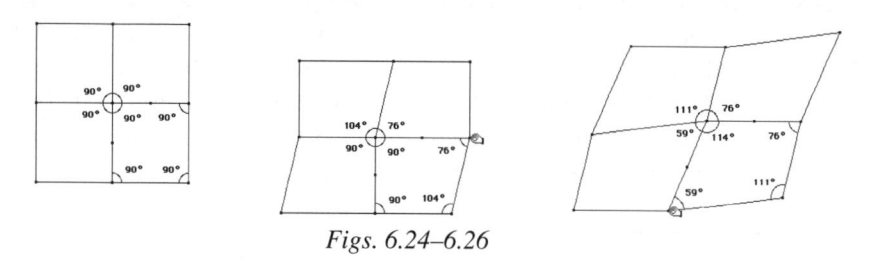

Figs. 6.24–6.26

Example 10 *Napoleon-Barlotti theorem for n = 4*
If we draw four squares about the sides of a base square and connect the centres of those four outer squares, a square is the result (Fig. 6.27). Modification of the base square into a parallelogram preserves this property (Fig. 6.28). However, this is only valid for parallelograms and cannot be generalised further. This is a specific case of the theorem of Napoleon-Barlotti (the case $n=4$): if regular n-gons are externally (internally) constructed about the sides of an n-gon, their centres form a regular n-gon if and only if the initial n-gon is affine-regular. (This problem was introduced in Chapter 2, Example 5.)

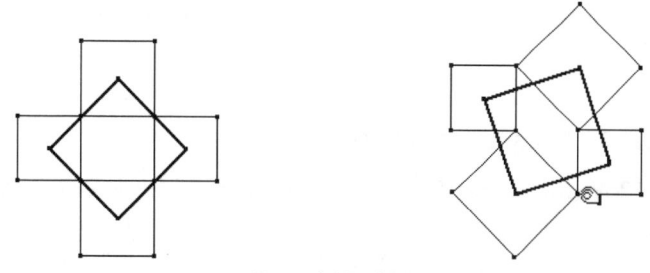

Figs. 6.27–6.28

Example 11 *Nested pedal-quadrilaterals [A harder example]*
A point is placed inside a square, initially at its centre (but not fixed there). Perpendiculars are dropped from the inside point to the four sides of the square and the four resulting intersection points (the 'feet') are joined to form a pedal-quadrilateral (a square actually) inside the original square. Starting from the base quadrilateral (which we call shape 1 – in our case a

square) the above process is iterated to produce four nested pedal-quadrilaterals (shapes 2-5) all with a common centre (Fig. 6.29, where the base quadrilateral is the outermost square). If the inside point is now moved away from its central position, by redefining it as a basic point, the innermost quadrilateral remains a square i.e. shape 5 remains similar to shape 1. Likewise shape 2 remains similar to shape 4 (Fig. 6.30). These similarity properties do not change even if the base square is transformed into a parallelogram (Fig. 6.31) or indeed into any quadrilateral (Fig. 6.32). In this way the following theorem is found (for the case $n=4$): the nth pedal-n-gon of an n-gon is always similar to the given n-gon.

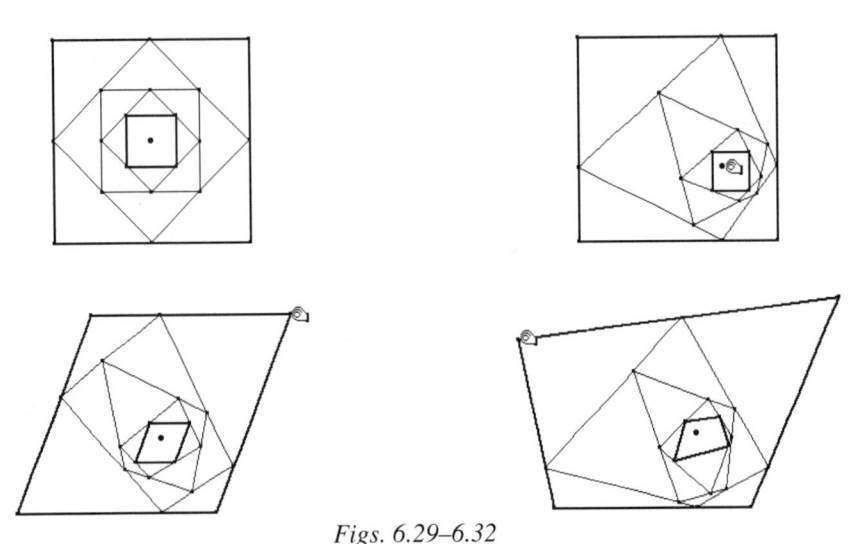

Figs. 6.29–6.32

6.2.3 Examples from hexagon theory

Example 12 *Napoleon-Barlotti theorem for $n = 6$ [A harder example]*
Six regular hexagons are constructed about the sides of a regular hexagon. Their centres form a regular hexagon (Fig. 6.33). If the base hexagon is transformed into an affine-regular hexagon, which is necessarily point-symmetric, by redefining certain dependent points as independent, the regularity of the centre-point hexagon is preserved (Fig. 6.34). This remains true even in the case of degeneration of the base hexagon to a line (Fig. 6.35) and in the case of 'internal' inversion of the hexagons drawn about the sides (Fig. 6.36).

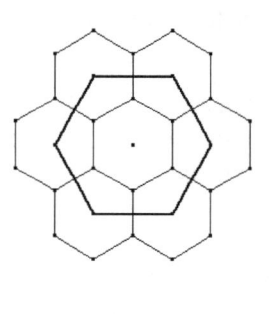

Figs. 6.33–6.34

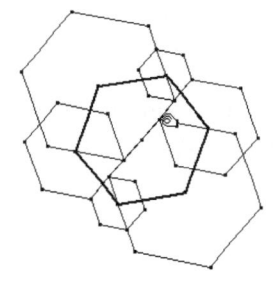

Figs. 6.35–6.36

Example 13 *A theorem about point-symmetric hexagons [A harder example]*

Starting from Fig. 6.33, the diagonals of the hexagon are drawn, creating two overlapping equilateral triangles (Fig. 6.37). We transform the initial regular hexagon into 'any' point-symmetric hexagon (N.B. not every point-symmetric hexagon is affine-regular – therefore two more pairs of opposite sides must each be parallel to a hexagon diagonal). Thus we obtain, trivially, a point-symmetric mid-point hexagon, which is usually irregular. But the regularity of the triangles which are formed out of its diagonals is invariant (Fig. 6.38).

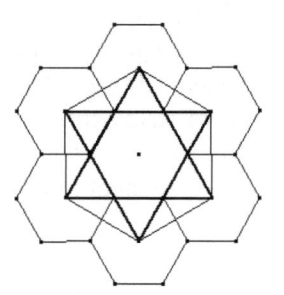

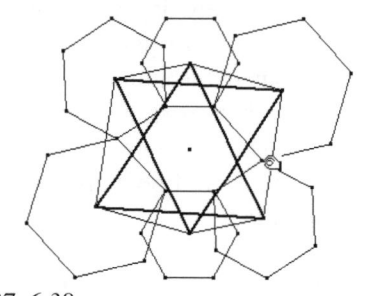

Figs. 6.37–6.38

6.2.4 An example of generalisation through redefinition of a straight line

In Sections 6.2.1–6.2.3 the generalisation has been achieved only by redefinition of a constructively dependent point as a basic point. Of course, other objects can be made into basic objects, as the next example shows.

Example 14
We wish to generalise the solution of the problem to construct an equilateral triangle with a vertex on each of two given parallel lines a, b with the third vertex at a given point P (Fig. 6.39: showing a double solution). In the initial configuration, a is constructed parallel to b; this means a is constructively dependent upon b. Through redefinition of a as a basic object, which need no longer be parallel to b, we obtain a generalisation of the problem with the corresponding generalised solution (Fig. 6.40), which we met in Chapter 2, Example 13.

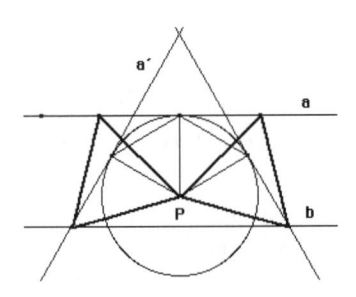

Figs. 6.39–6.40

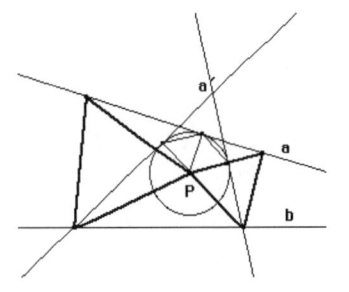

6.3 FINAL REMARK

The method of theorem finding through conceptual generalisation by redefining constructively dependent objects as basic objects is only appropriate to certain theorems of plane geometry. For example, no generalisation can be carried out if the number of intervening geometric objects has to be changed – such as with analogous generalisations of statements about triangles to statements about quadrilaterals. For this reason this method of generalisation should be considered and applied in conjunction with other heuristic methods.

CHAPTER 7

Producing and using loci

*The Moving Finger writes,
and having writ, moves on.*
Edward Fitzgerald

7.1 INTRODUCTION

The heuristic and experimental learning of plane elementary geometry can be strengthened with the use of suitable 2–D graphics systems. In particular, cinematic observation ('moving pictures'), which until now depended on the students' (and the teacher's!) powers of visual imagination, is now a reality. In this chapter we explore the interactive production of loci, which is a feature distinguishing Cabri-géomètre from many other systems for geometric construction in schools.

Description of the method
- A construction is carried out. (In our example this is the construction of the medians of a triangle, shown in Fig. 7.1).
- The points which determine the construction, the so-called basic points, are free to move. (In our example these are the three vertices of the triangle).
- A basic point or other basic object is connected to a specially selected path using **Redefine an object** (or **Link a point to an object**) from the **Miscellaneous** menu, if it is not already connected to such a path by the construction itself. (In our example B is the basic point and the circumcircle of the triangle is the path, as shown in Fig 7.2).
- The movement of the basic point along the chosen path results in variation of the constructed configuration. (In Fig. 7.3 the 'gripping hand' icon indicates the cursor dragging B.) During this transition along the path, we are only interested in the movement of one selected point,

whose construction depends on the basic point. (In our example this is the centroid where the medians intersect, which moves as basic point B moves.)

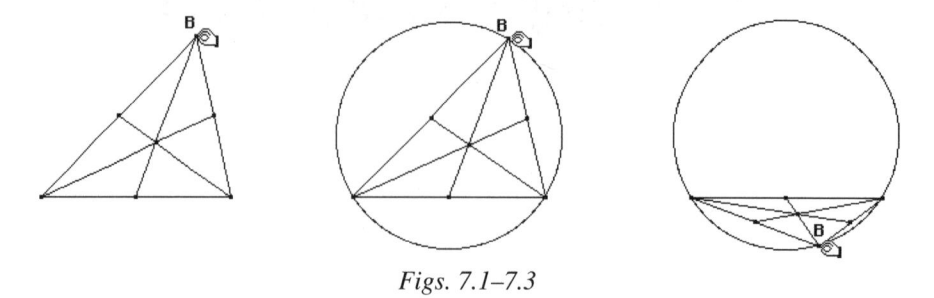

Figs. 7.1–7.3

The question of interest is: Which locus (path a) is described by point A when point B is moved on its guide-path (path b)? We prepare to move B in order to observe the path followed by A (Fig. 7.4). To generate this locus point-by-point, the **Construction** menu option **Locus of a point** is chosen; Point A is marked (arrow symbol in Fig. 7.5) and point B is moved along its path (using the mouse to drag it); A leaves behind a trace of **discrete** points on the screen, which must be associated with the shape of the locus a by the user (in the example it is a circle, Fig. 7.6). The set of points of a certain path is mapped (mostly bijectively or 1–1) onto a set of points forming an image path a. In the given example, the circumcircle is mapped onto the circular path followed by the centroid. Normally, the concept of function comes into geometry in the context of volume or area or length measurements but the context of loci offers a new possibility of introducing functional dependencies (one path being a function of another).

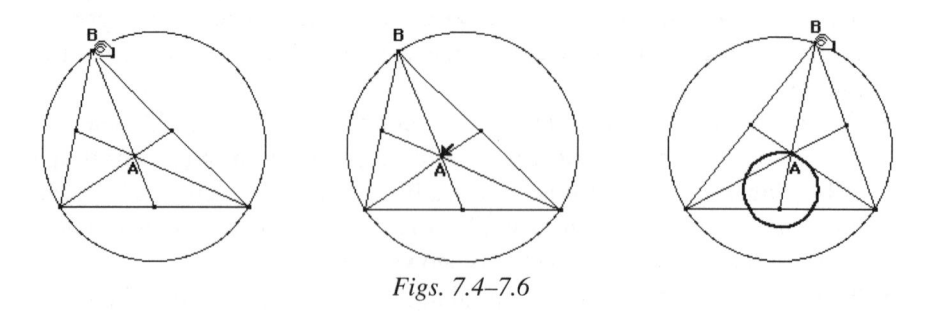

Figs. 7.4–7.6

Studying loci using such a system can be effectively applied in mathematics teaching in the following contexts:

Producing and using loci

- in the heuristic phase of construction task problem solving;
- in the experimental verification of construction results;
- in investigations of the position and shape of the image of a transformed original shape;
- in the construction of conic sections and algebraic curves of order >2;
- in investigations of the shapes of loci generated by the movement of special points in a triangle.

Using software eliminates problems of lack of time, motivation and accuracy, which inevitably arise using conventional tools, and which have led to widespread avoidance of loci in school geometry in the past.

7.2 APPLICATIONS OF LOCI

The possibilities are explored using selected examples, in which heuristic strategies come in useful (the corresponding synthetic or analytical proofs are not the subject of this book). The static screen dumps presented can only inadequately document the dynamic processes in the production of elementary loci. The following examples are designed to stimulate your own work with this new tool and explore its relevance to elementary geometry – personal experience is essential for it to be properly assessed.

7.2.1 Loci as an aid to finding solutions to construction tasks

The interactive production of loci is an important component of a total treatment of interactive construction tasks. Using loci is particularly valuable for solving a certain class of transformational geometry problems known as insertion problems. We begin with a classic example.

Example 1

A square is to be inscribed in a triangle such that two vertices of the square lie on one side of the triangle and the other vertices of the square lie one on each of the other two sides of the triangle. (This problem was tackled in a different manner in Chapter 6, Example 7.)

First we satisfy the required position of three of the square's vertices (i.e. leaving out one condition) and we draw with the aid of a macro a suitable square (Fig. 7.7), and then vary it. Along what path does the remaining free corner of the square move? Its locus, together with an experimental solution, are shown in Fig. 7.8; thus the construction method is discovered.

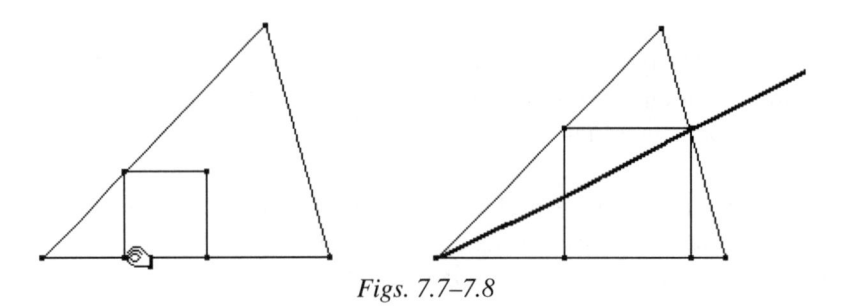

Figs. 7.7–7.8

Example 2
In this example three parallel lines are given. We seek an equilateral triangle whose vertices lie one on each of these lines.

In Fig. 7.9, an equilateral triangle is drawn (using a macro, for instance) with B and C as basic points and A the dependent third point, while ignoring one condition. We move B and generate the locus of A (Fig. 7.10, showing an experimental solution). It is found that whatever the independent positions of the three parallel lines, the locus makes the same angle with these lines, namely 60°.

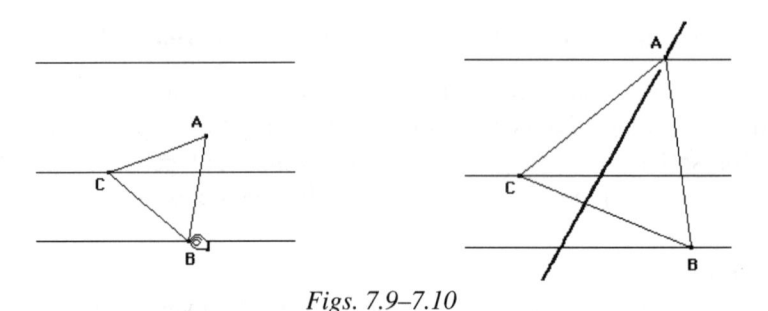

Figs. 7.9–7.10

Example 3
We are given a circle k, a straight line g and a point P. We seek a line segment with midpoint P whose endpoints lie on the circle and the line.

First we satisfy the requirement that one of the endpoints, K, lies on the circle, and reflect this circle point in P to give K' (Fig. 7.11). We generate the

Producing and using loci

locus of K' as K is moved on the circle k, and obtain as locus a circle (which must be the reflection of k in P) whose points of intersection with line g are the required endpoints (Fig. 7.12, showing an experimental solution) Now, by changing the position of P we can investigate the general solubility problem (Fig. 7.13).

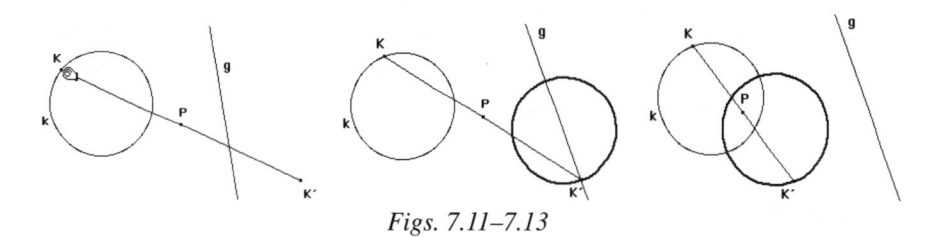

Figs. 7.11–7.13

Example 4 *[A harder example]*

In order to give an example from a conventional setting, we turn back to Section 7.1 (Fig. 7.6). There, a circle had been produced as the locus of the centroid when a vertex of a triangle was moved along its circumcircle. Where does the centre of this centroid-circle lie? What is its radius? We recognise that the centroid-circle lies symmetrical to the perpendicular bisector of the unchanging side of the triangle, and construct a circle through the centroid A with moveable centre lying on this perpendicular bisector (Fig. 7.14). We keep moving the centre until this circle covers the centroid-circle (Fig. 7.15, compare with Fig. 7.6). By measuring we find: the radius of the circle equals one third the radius of the circumcircle and the midpoint is one third the way along the perpendicular line from the circumcentre along the unchanging side of the triangle.

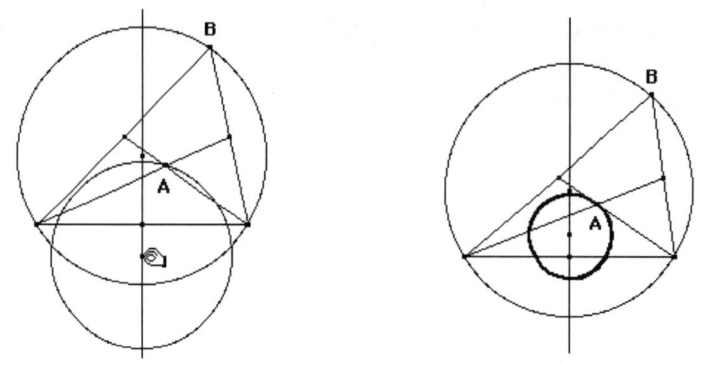

Figs. 7.14–7.15

7.2.2 Experimental checking of construction results

Suppose we have constructed a figure using line segment, line or circle as basic objects, as is often the case. If the same figure can also be produced point by point as a locus, an opportunity to check the result arises: Does the object constructed as a whole correspond with its locus? We give some examples for such a test, useful for checking whether mistakes have been made in a construction. In addition, it can give information about further properties of a constructed object as well as helping to reinforce geometric knowledge. From a logical viewpoint, such a test does not, of course, suffice to prove the correctness of a construction.

Example 5

Following up section 7.2.1, we construct the centroid-circle of a triangle simulating the standard ruler-and-compasses method (Fig. 7.16). The locus of the centroid verifies the construction result (Fig. 7.17).

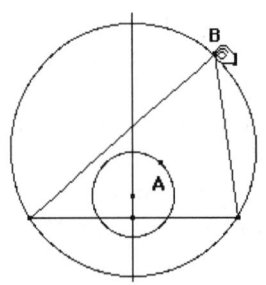

Figs. 7.16–7.17

Example 6

Using a macro for an enlargement about O, the two points M and P have images M' and P' (Fig. 7.18). Is the circle passing through P' with centre M' really the image of the circle passing through P with centre M after the enlargement? The answer is yes.

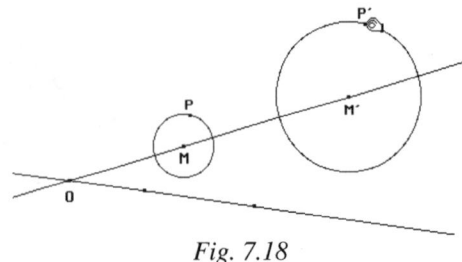

Fig. 7.18

Producing and using loci

Example 7

Two straight lines meet at a vertex to produce an angle. The angle bisector is generated with a macro (Fig. 7.19). It is then checked that the angle bisector has the following properties:

- It is an axis of reflective symmetry for the two line configuration, i.e. a reflection in it reproduces one of the given lines on top of the other (Fig. 7.19).
- It consists of the midpoints of all pairs of points (one on each of the two lines) which are equidistant from the vertex (Fig. 7.20).
- Its points have the same perpendicular distance from the two straight lines (Fig. 7.21).

We can use the locus facility to check these properties.

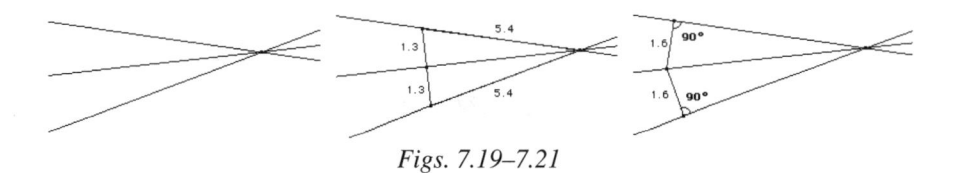

Figs. 7.19–7.21

Example 8 *[A harder example]*

We are interested to know what is the image of a small circle centre M radius MN under circle inversion in the large circle with centre O (Fig. 7.22). As a first attempt we find the image M' of the centre M and the image P' of a variable circumference point P and infer that the circle's image is the circle centre M' which passes through P'. (This strategy worked in Example 6.) Using the **Locus** command we see that our assumption was wrong! (The image of the centre is *not* the centre of the image – Fig. 7.23.)

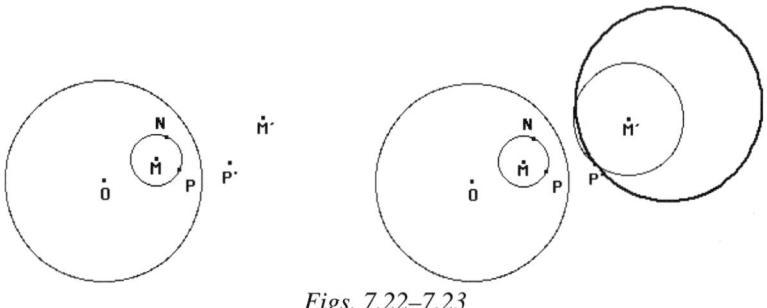

Figs. 7.22–7.23

125

7.2.3 Investigating image–figures (as sets of points)

In the following examples, the images of circles are generated. In Figs. 7.24 and 7.25, the circle image is produced by a line reflection; in Figs. 7.26 and 7.27, a 'screen pantograph' is used for enlarging a circle; in Figs. 7.28 and 7.29, an ellipse is produced as the image of a circle under an axis-affinity transformation: (g is the fixed axis, P maps onto P' by a one-way stretch with stretch factor -2.)

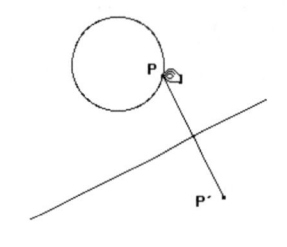

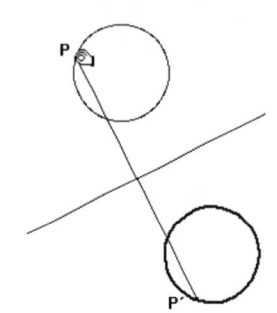

Figs. 7.24–7.25

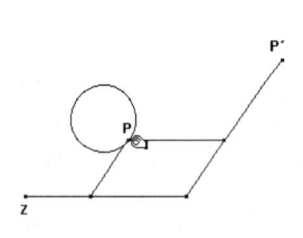

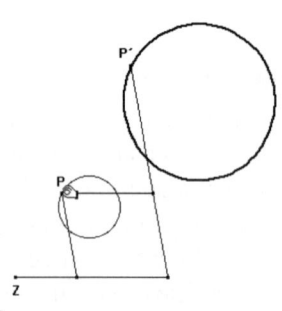

Figs. 7.26–7.27

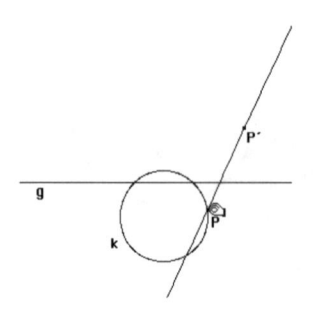

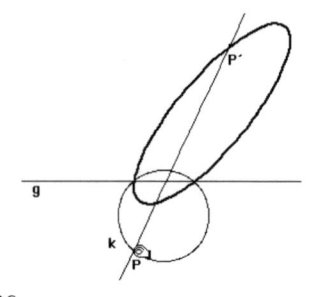

Figs. 7.28–7.29

Figs. 7.30–7.32 show the point by point generated images of the circle under a central projection (centre Z, disappearance line v, axis s). The nature of the circle's image depends upon whether the circle misses, touches or intersects the disappearance line v, producing respectively an ellipse, parabola or hyperbola.

Figs. 7.30–7.32

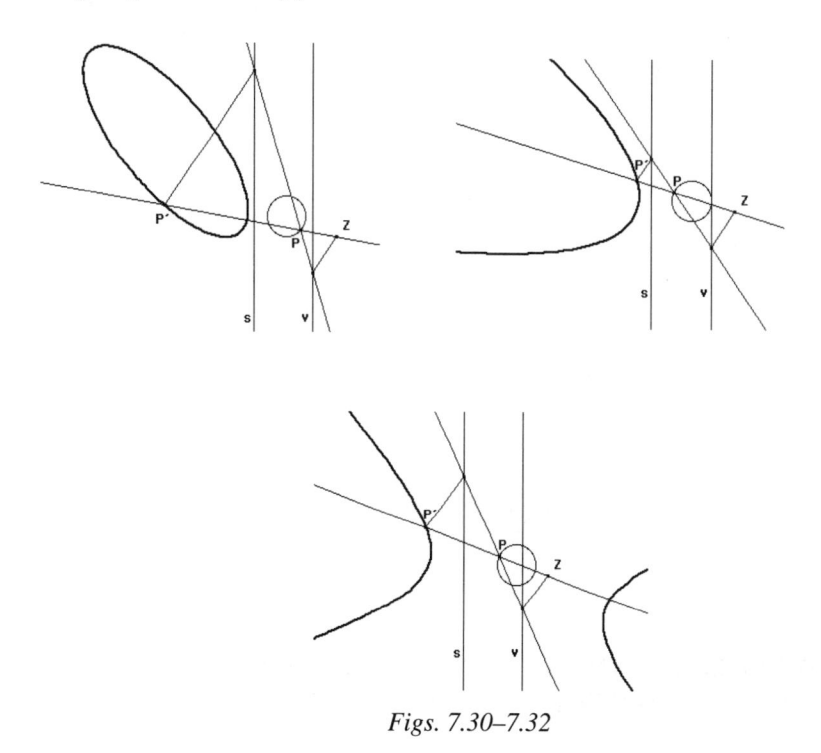

Figs. 7.33–7.35 represent the images of circles under an inversion of the circle (we limit ourselves here to circles which lie within the inversion circle). If the circle touches the inversion circle then so does the image circle, at a common point (Fig. 7.34). If the circle passes through the centre of the inversion circle, its image is a straight line – i.e. a circle of infinite radius (Fig. 7.35). With this result, a circular movement can be converted into a straight line movement: this produces the construction of Peaucellier's Cell (Fig. 7.36; dotted lines indicate construction lines normally hidden as in Fig. 7.37). (A fuller explanation is to be found in Chapter 10, Example 10.) We test our screen Peaucellier Cell, and *do* obtain a straight line for the locus of P' as P moves round the circle (Fig. 7.38).

Figs. 7.33–7.35

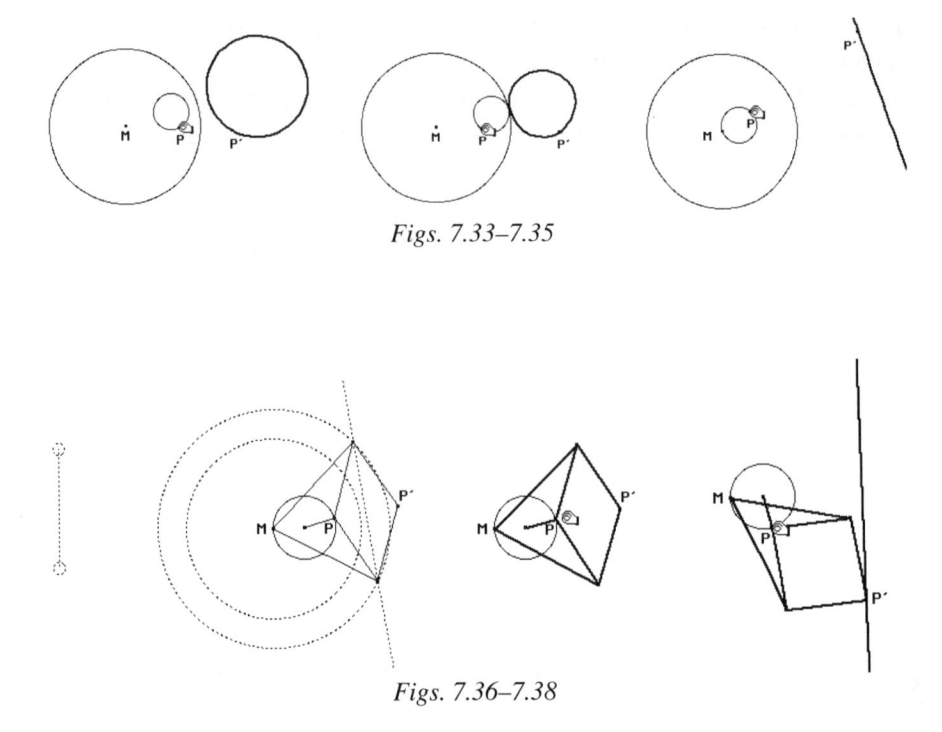

Figs. 7.36–7.38

7.3 ADVANCED USE OF LOCI: CONSTRUCTION OF CONIC SECTIONS AND OTHER ALGEBRAIC CURVES

In this section we consider some more advanced uses of loci to construct curves. The point by point construction of curves is very time consuming which is certainly an important reason why few curves are produced in this way in elementary geometry teaching. However, with interactive generation, only a single application of the relevant construction is needed to construct curve point A which depends upon a base point B. As B is moved along its path, A passes in turn through all the points of the curve to be generated.

7.3.1 Generation of conic sections

In this section, we have compiled a collection of common conic section constructions, which can be further augmented by those illustrated in Figs. 7.28–7.32 of Section 7.2.3.

Producing and using loci

Figs. 7.39–7.40 show the construction of an ellipse through perpendicular-axis affine imaging of the main circle of the ellipse (as is often done in the introductory school example, A is chosen as the bisection point of the perpendicular line from the axis to B). Figs. 7.41–7.42 show the vertex–circle construction of the ellipse.

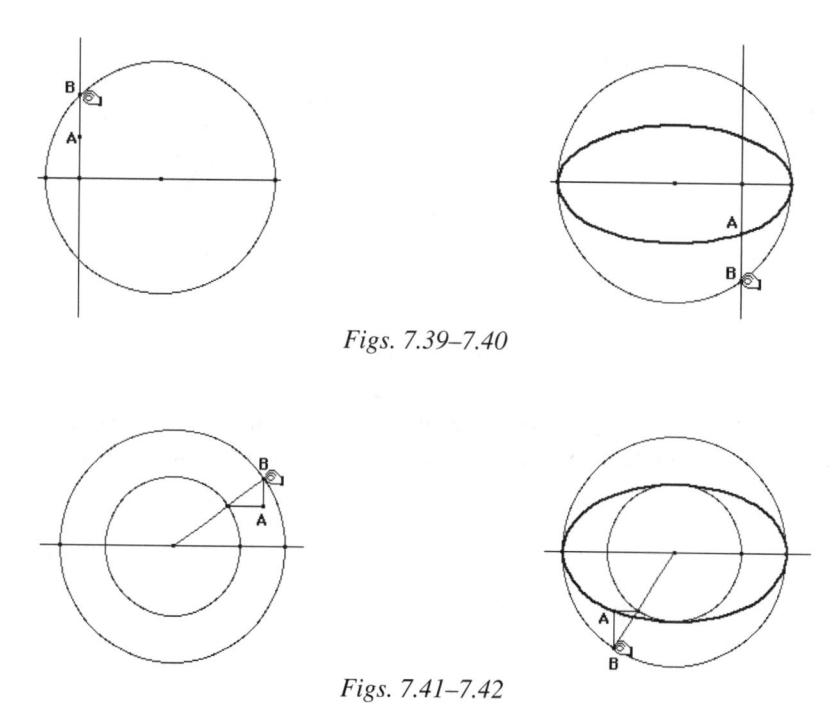

Figs. 7.39–7.40

Figs. 7.41–7.42

Other constructions include the tangent to an ellipse. Thus in Figs. 7.43–7.44, the guide-circle of the ellipse is shown. The guide-circle centred about the focal point F_2 has the distance of the major axis of the ellipse as its diameter. The perpendicular bisector of BF_1 is the tangent of the ellipse through A, the point of intersection of the perpendicular bisector and BF_2.

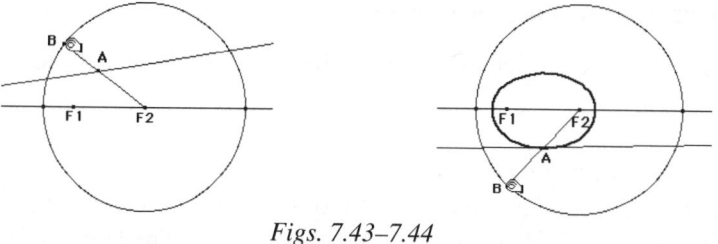

Figs. 7.43–7.44

129

Discovering Geometry with a Computer

The guide-circle construction of the hyperbola is analogous (Figs. 7.45–7.46).

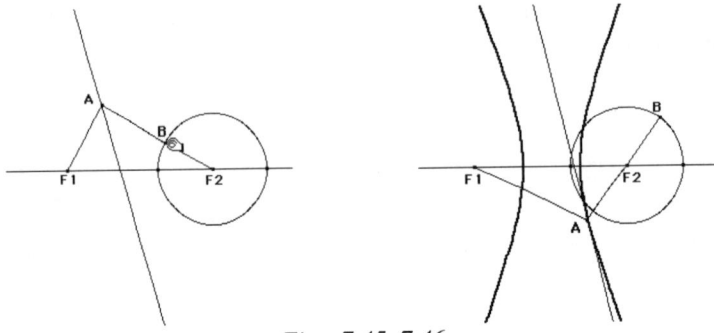

Figs. 7.45–7.46

In the guide-circle construction of the parabola (guide-path = guide-circle with 'infinitely large' radius), the perpendicular bisector of FB is the tangent through A of the parabola, the distance of each parabola point to the guide–line, l, equals its distance to the focal point (Figs. 7.47–7.48).

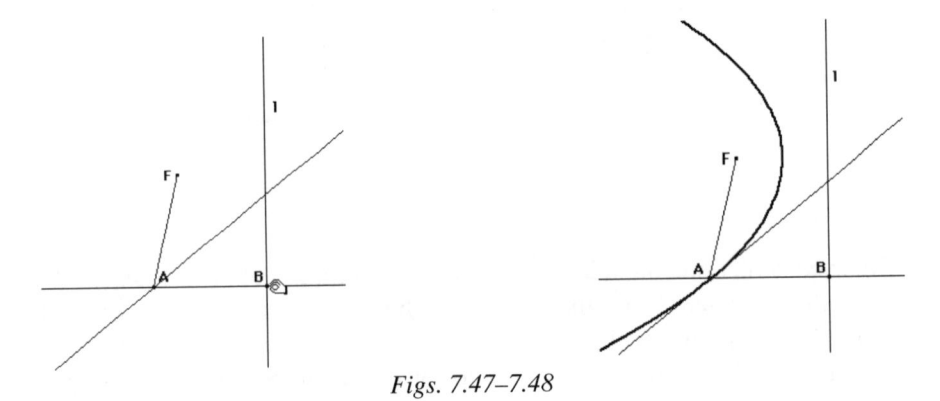

Figs. 7.47–7.48

7.3.2 Generation of algebraic curves

Cabri-géomètre is particularly suitable for the cissoidal generation of algebraic curves.

General definition of a cissoid
Curves k_1 and k_2 and a point P (centre of rotation) are given (Fig. 7.49). A straight line through P intersects k_1 at B and k_2 at G. The length of the vari-

Producing and using loci

able line segment BG on this straight line is copied on both sides of P to produce PA_1 and PA_2. What is the locus described by A_1 (or A_2) when B moves along k_1? (Curves thus produced are called cissoids.) Since Cabri-géomètre only allows straight lines and circles for k_1 and k_2, only special algebraic curves up to the fourth order can be produced. There are several cases which can be investigated.

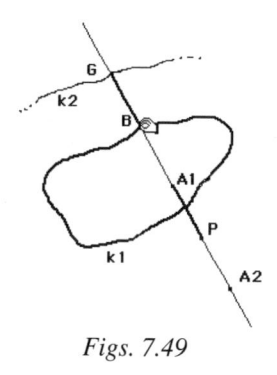

Figs. 7.49

Example 9

We consider an historically important case of the cissoid: circle as k_1, straight line as k_2, initially P lies on k_1 and the perpendicular from P onto k_2 passes through M, the centre of the circle (Figs. 7.50–7.51). PA is constructed here in the same direction as BG, using an appropriate macro (Figs. 7.50–61 present a series of results for different positions of A).

[Note: initially we choose A_1 to be the point A; later we will choose A_2 to be A.]

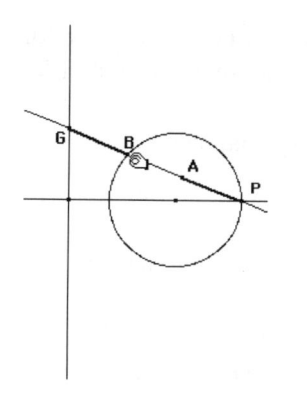

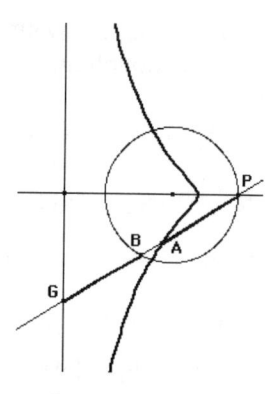

Figs. 7.50–7.51

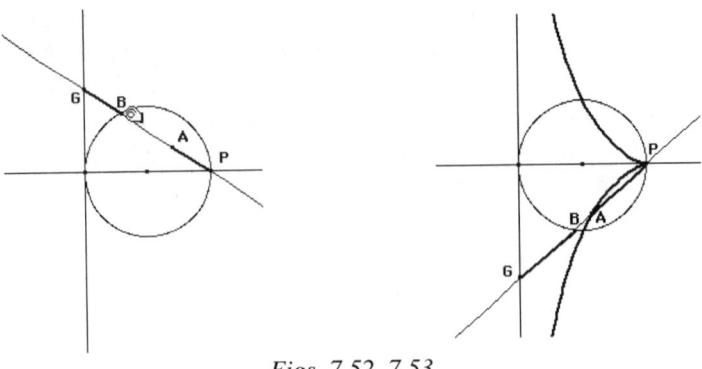

Figs. 7.52–7.53

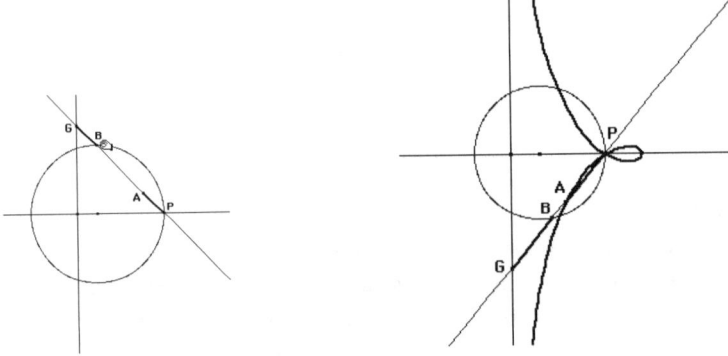

Figs. 7.54–7.55

Our method is specialising: we move the line k_2 in stages from the 'left side' to the 'right side' of the circle k_1. For each stage we generate a new cissoid and observe its form:

- Fig. 7.53 shows the cissoid of Diocles which was used in antiquity for cube–doubling; it is the 'oldest' algebraic curve other than those of order 2.
- Fig. 7.55 shows an intermediate cissoid without special name.

Now we take A to be A_2 in Fig. 7.56 and thus:

- Fig. 7.57 shows Newton's strophoid (line k_2 passes through the centre of circle k_1).
- Fig. 7.59 shows Maclaurin's trisectrix which was used for angle trisection (line k_2 passes through the midpoint of the radius of circle k_1).
- Fig. 7.61 shows another nice example without special name.

Producing and using loci

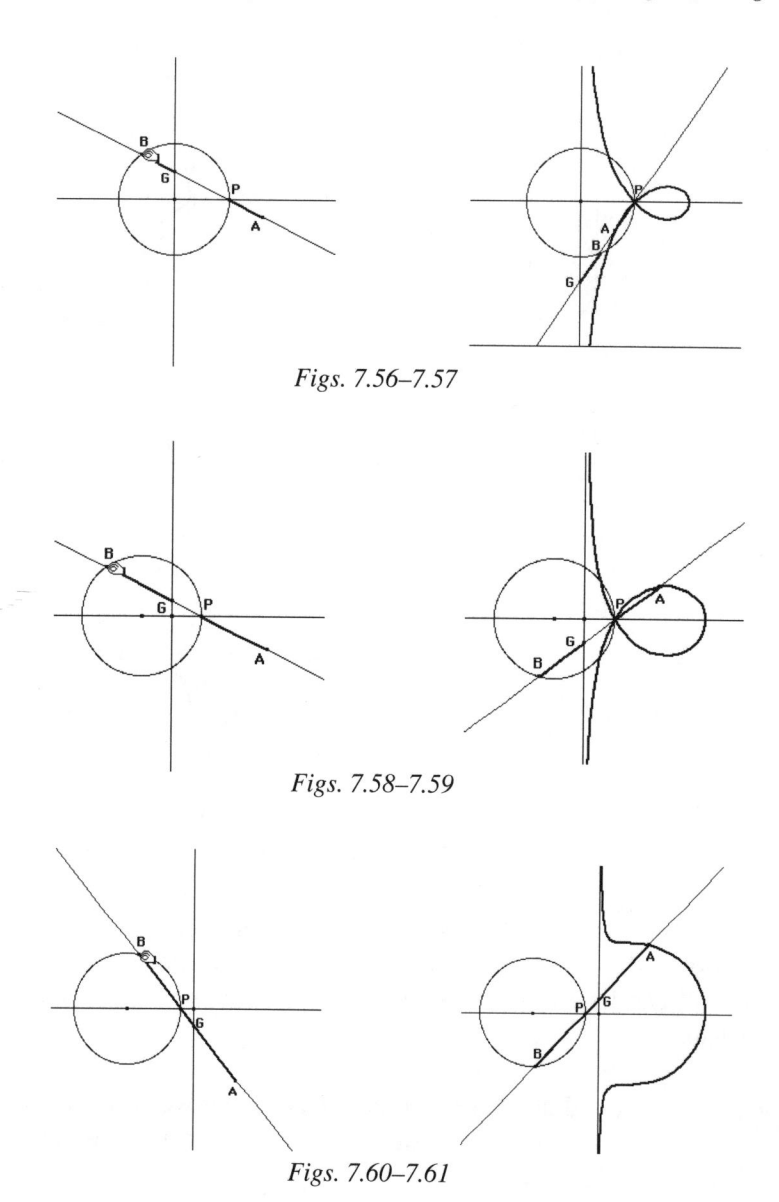

Figs. 7.56–7.57

Figs. 7.58–7.59

Figs. 7.60–7.61

Finally we can vary the position of the centre of rotation P (the pole) on the perpendicular. An example is given in Figs. 7.62–7.63; this algebraic curve falls into two branches, which have k_2 as asymptote.

In Figs. 7.64–7.65 Pascal's limaçon is produced cissoidally employing two circles as k_1 and k_2.

Figs. 7.62–7.63

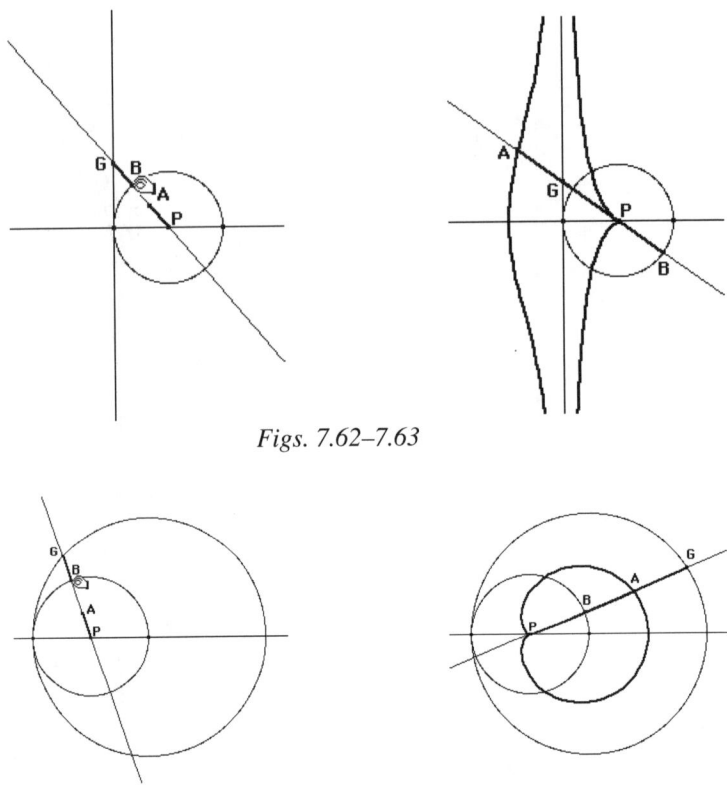

Figs. 7.64–7.65

The algebraic curves known as conchoids can also be produced. The production of pedal curves (as illustrated above) of conic sections is simple (which we can link with the last three constructions of Section 7.3.1; see also Chapter 10).

Computer supported generation of loci greatly improves access to the aesthetically attractive world of forming algebraic curves above order 2.

7.4 INVESTIGATIONS OF LOCI OF SPECIAL POINTS IN THE TRIANGLE

We end our list of applications – which is by no means complete – with a triangle geometry exercise. Starting from the introductory example in this chapter (see Figs. 7.1–7.6), we further examine which loci are described by particular points of the triangle if one vertex is moved on a straight line or on a circle. Thus the orthocentre of triangle ABC describes a circle (the 'orthocircle') if C is moved round the circumcircle; that circular locus is in fact the image of the circumcircle after reflection in AB (Fig. 7.66).

The centre of the incircle of ABC describes an arc with A and B as its endpoints as C moves along the circumcircle (Fig. 7.67). What are the centre and radius of this circular arc? What is the locus if C goes 'below' AB and the triangle is inverted?

As C moves along the circumcircle of ABC, the centre F of the Nine-point circle describes a circle congruent to itself, whose centre coincides with the midpoint of AB (Fig. 7.68).

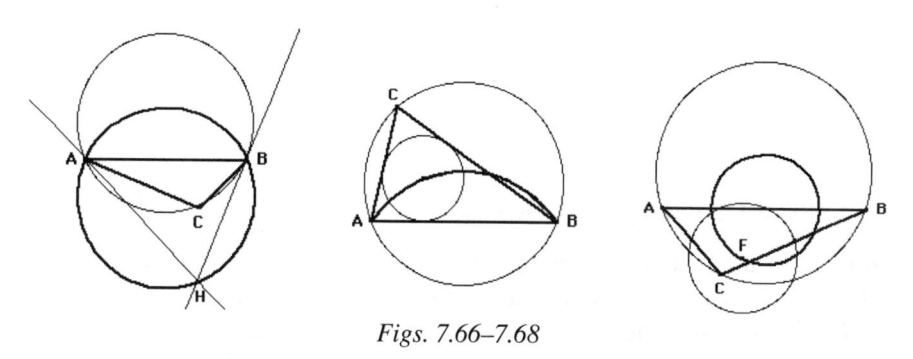

Figs. 7.66–7.68

In Fig. 7.69, let K be the point of intersection of those lines parallel to the bisectors of the interior angles of the triangle which pass through the midpoints of the respective opposite sides. We obtain an attractive curve, symmetric about the perpendicular bisector of AB. What is the description of its parameters?

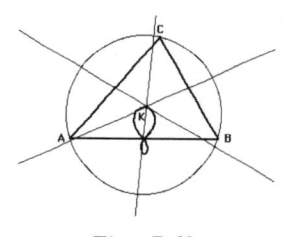

Figs. 7.69

And so we may continue ...

If we amend the above configuration so that only two of the three vertices lie on a circle, we obtain further curves such as shown in Fig 7.70. A further interesting locus for this figure is obtained by the path of the orthocentre as one vertex of the triangle moves on a straight line. The reader might like to speculate what the path will be ... (In Fig. 7.71, the locus has been drawn automatically, not by hand, a facility available in recent versions of the software.)

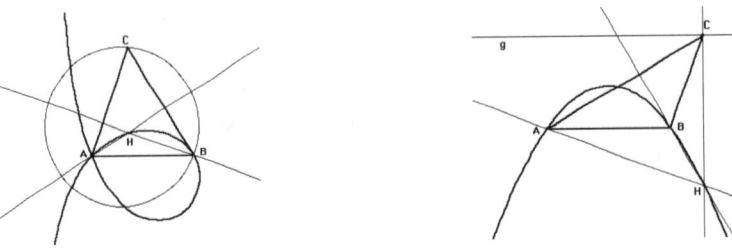

Figs. 7.70–7.71

7.5 CONCLUDING REMARKS

Note 1: In this chapter we have restricted ourselves to the production of just a single locus dependent on a point which is moved. It is, however, possible to produce the loci of several (up to 12) constructively dependent points *at the same time*; thus loci of loci can be produced (see Chapter 10, Example 8b). In addition, the independently moved point can draw its own trail; thus freehand lines and their images can be generated (see Chapter 8, Section 8.5).

Note 2: In cases where no direct construction is available for the production of a particular curve, as with algebraic curves for instance, we can use the following explorative scheme.

I. Choose a point and generate its locus by moving some other basic point.

IIa. Is an interpretable locus produced? YES!
We have a recognisable geometric theorem (possibly after further variation of the configuration). The investigation is completed! Here is an example.

Example 10 *Interpretable locus*
The point of intersection of the perpendicular bisectors of two sides of a triangle, AC and BC, is constructed (Fig. 7.72). Free variation of C shows that the point of intersection generates a straight line, which in fact is the third perpendicular bisector (Fig. 7.73). This indicates the theorem that all three perpendicular bisectors always meet at one point.

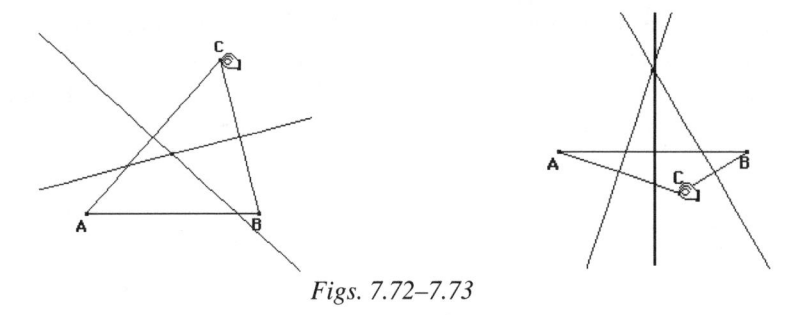

Figs. 7.72–7.73

IIb. Is an interpretable locus produced? NO! Here is an example.

Example 11 *Uninterpretable locus*
Fig. 7.74 shows two altitudes and the orthocentre H of triangle ABC. Fig. 7.75 shows various positions of H as C is freely moved. Nothing new is learnt here – the positions of H give no special shape – just a mass of dots. The investigation fails!

We then proceed to stage **III** in order to try to complete the investigation with a positive discovery.

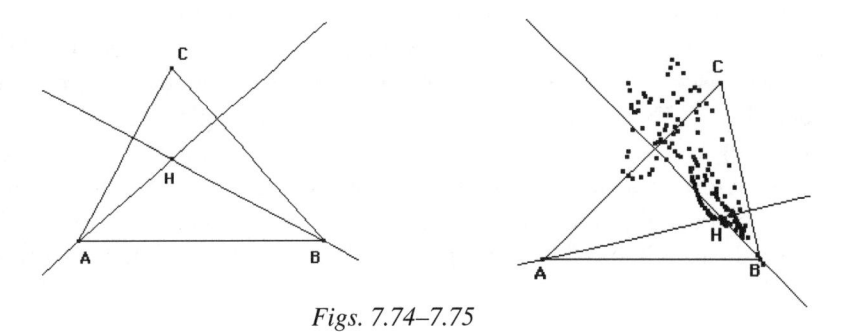

Figs. 7.74–7.75

Discovering Geometry with a Computer

III. We link the free point to some particular path (circle, line, or line segment) to constrain the motion of the free point. This may produce a recognisable locus. Here are two examples.

Example 12 *Constraining a point to produce a locus – a parabola*
Starting from Fig. 7.74 again, we draw in the third altitude and constrain C to lie on a basic line g by placing a point on g and using **Redefine an object** (or **Link a point to an object**) to identify C with the point on g (Fig. 7.76). The locus now produced by H as C moves along g is as shown in Fig. 7.77. This we can just recognise as a parabola. (Where are the focus and directrix of this parabola?) The investigation is completed successfully!

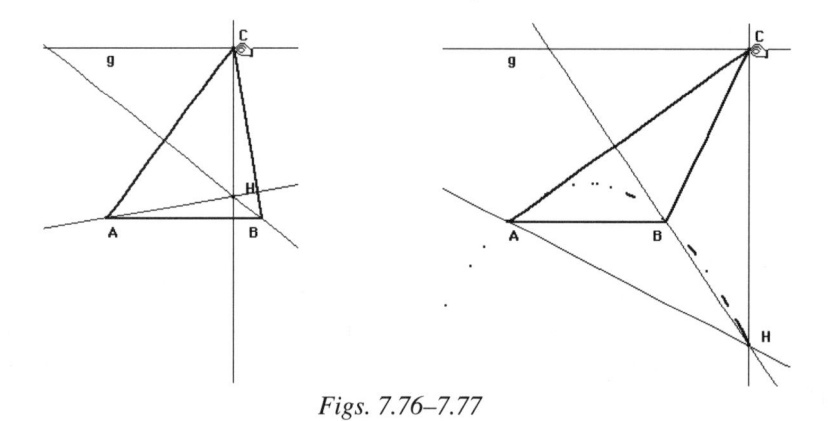

Figs. 7.76–7.77

Example 13 *Constraining a point to produce a locus – a straight line*
While carrying out three consecutive reflections in straight lines (possibly using a macro), we investigate what locus is described by the midpoint of the line from point P to its associated image P' as P is freely moved. There are two cases to consider, depending on whether the three lines are coincident (Fig. 7.78 is an example) or enclose a triangle (Fig. 7.80 is an example). We see that in both cases, with free movement of P, the midpoint of PP' describes a straight line. Following investigation of further cases we arrive at the well known theorem concerning three reflections (Figs. 7.78–7.79 indicate a line reflection; Figs. 7.80–7.81 indicate a glide reflection).

Producing and using loci

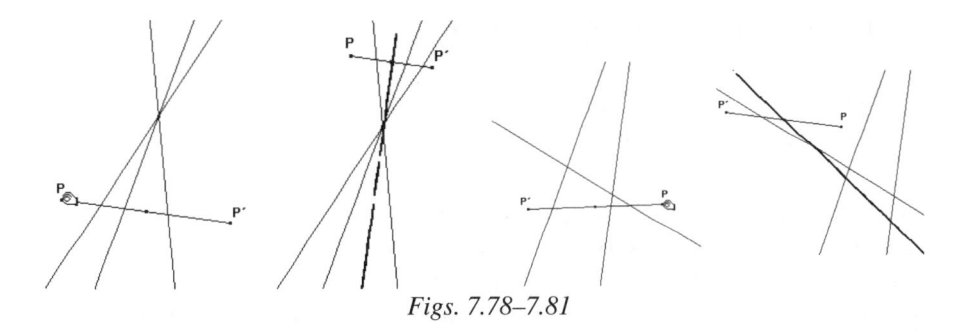

Figs. 7.78–7.81

Note 3: The quality of a point-by-point generated locus depends on the skill and dexterity of the user of the mouse. It is almost impossible to watch simultaneously both the independently moved point and the dependent point which generates the locus. One instinctively watches the dependent point (probably because of the suspense and curiosity as to what sort of curve is being produced). The density and distribution of the points of the resulting locus can be very variable (compare the loci drawn in Figs. 7.71 and 7.77). Recent versions of Cabri-géomètre allow the automatic drawing of loci, in those cases where there is a guide-path, which is a considerable improvement.

In Fig. 7.77, the points may not be interpretable as a meaningful whole unless the observer has previous relevant experience of curves. Recognition of the locus as an entity (or gestalt), despite the gaps between its individually produced points, is explained in the psychology of perception as the formation of a 'bridge'. The recognition of a bridge is instinctive, in contrast to the case where a line is *wilfully* constructed in the mind from observing a set of discrete points. Research is needed to establish under what conditions of density and distribution of points students can recognise loci. This will depend on the type of locus and the ages, abilities and knowledge of the students. Their understanding might be measured by asking them to draw pencil sketches of loci which appear on screen.

The question arises, whether these psychological perception problems should be avoided by having a drawing system which simply produces the locus automatically, as alluded to above. The following arguments speak against using automatic generation (if available) and in favour of at least the option of retention of the present generation mode:
– Training in gestalt perception can be enhanced.
– The student is actively involved throughout the generation process.

- There may be value in retaining some tactile dynamic activity in curve generation.
- The speed of the hand movement and the speed of the drawn locus can be related, which may lead to significant insight.

Note 4: At present with Cabri-géomètre it is not possible to further process a generated locus. The locus is only a state of the screen and the system has no knowledge of its nature – it is merely composed of individual pixel points with no internally recorded structure. Further processing of the locus by computer is conceivable in two principle ways (but not available in Cabri-géomètre or any other package as yet):

- Numeric solution: Curve fitting to (some of) the locus points (e.g. spline interpolation).
- Algebraic solution: Derivation of an exact analytic–geometric representation of the locus, using an expert system.

The second solution is at present not possible because formidable deduction problems can arise in the analytical description of loci (e.g. transcendental curves). With regard to use with secondary school students, one would limit oneself to conic sections to match the current curriculum. J-M Laborde (1988) has described a first attempt at this.

Note 5: Not all loci can be produced with the current version of Cabri-géomètre. The production of envelopes and cycloids is impossible except in certain special cases (see the book cover for an example). The reasons for these restrictions lie in the type of transformation of geometric configurations possible in drag-mode. There is a real danger in such restrictions: we only cover those constructions which are feasible with the given tool – an instance of 'tool-conditioned selectionism'.

Note 6: With the interactive generation of a locus, the position of only one locus point actually need be constructed, the rest being generated as the independent point is moved. This is in contrast to the traditional ruler-and-compasses method, which Cabri-géomètre simulates, where the same construction must be repeated point by point, in order to have sufficient points to connect by freehand interpolation to produce a continuous curve. Computer-supported generation certainly has the advantages of time-economy, modification possibilities, etc. It makes exploring considerably easier, but the effects of practice and internalisation of the conventional construction of loci are partly lost – the students have figures produced for them and do not construct them for themselves any more. To draw loci using Cabri-géomètre, and to do so using conventional tools on paper involves quite different experiences and skills. It is quite wrong to discredit the traditional approach – both it and computer-based methods are valuable. The choice of tool depends on the objectives.

CHAPTER 8

Generating geometrical figures with line symmetry

*God hath given you one face,
and you make yourself another.*
Hamlet

8.1 INTRODUCTION

The construction of symmetric figures using ruler and compasses is slow, inaccurate and cannot be repeated or varied easily. Interactive tools such as Cabri-géomètre overcome these problems and furthermore offer wider construction possibilities. With such systems it is possible to transform a screen figure with no symmetry into one having symmetry and transform one of low order symmetry into one of higher order symmetry. With Cabri-géomètre, generation of symmetric figures is made quick and simple by using macros for producing figures (symmetric or not) and using further macros to produce reflected images of such figures. The generation of free-hand drawn lines, which Cabri-géomètre allows, extends the possibilities much further – into individual and creative forms.

In this chapter we restrict ourselves to the interactive generation of line-symmetric figures on a plane. (One can generate any translation or rotation by combining two reflections.) The following examples demonstrate the many possibilities of Cabri-géomètre.

8.2 AN EXAMPLE OF THE GENERATION OF LINE-SYMMETRIC FIGURES

We construct a well-known figure out of seven circles (Fig. 8.1) which is determined by the centre and a circumference point of a base circle. Fig. 8.2 shows the macro definition with the initial two objects marked by small cir-

Generating geometrical figures with line symmetry

cles and the target objects marked by dotted circles. After clicking on the macro 7-CIRCLES (Fig. 8.3) and two points, a drawing like that in Fig. 8.1 is constructed on which we can superimpose the axes of symmetry (Fig. 8.4).

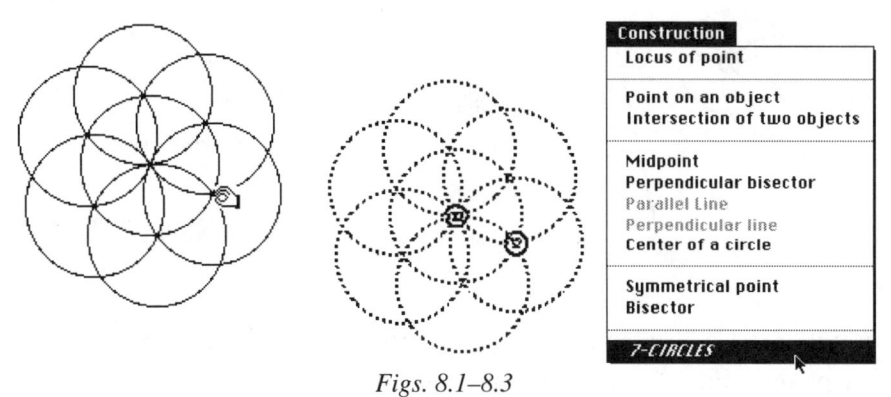

Figs. 8.1–8.3

We can check the 6-way axial symmetry by reflecting a circle point in an axis and then moving the point on the circle – its 'reflection points' run through the 'reflection circle' (not illustrated). The size and position of the figure with its axes of symmetry can be varied (Fig. 8.5). Obtaining the reflection in a line of the seven circle figure (without the axes of symmetry drawn in) is simple. We draw in the required reflection line and use the **Construction** option **Symmetrical point** twice, each time specifying one of the two points which determine the seven circle figure. Then we apply the macro 7-CIRCLES to the two image points. Figs. 8.6 shows the definition of the macro 7-CIRCLES-REFLECTED: the initial objects are now two points (shown as small bold circles) and a straight line (shown bold), the target objects are the seven circles of the original figure and those of the reflected figure and the points which determine them (shown dotted).

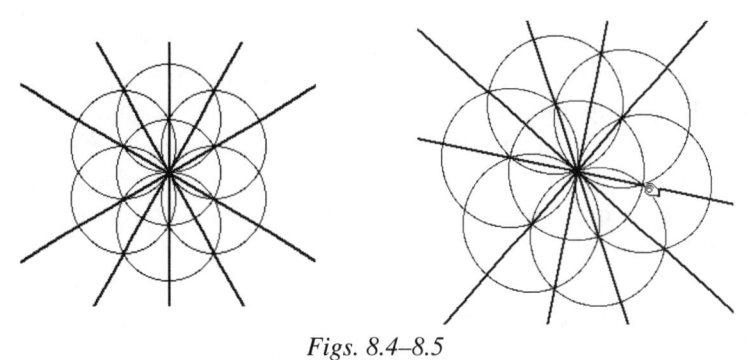

Figs. 8.4–8.5

143

Discovering Geometry with a Computer

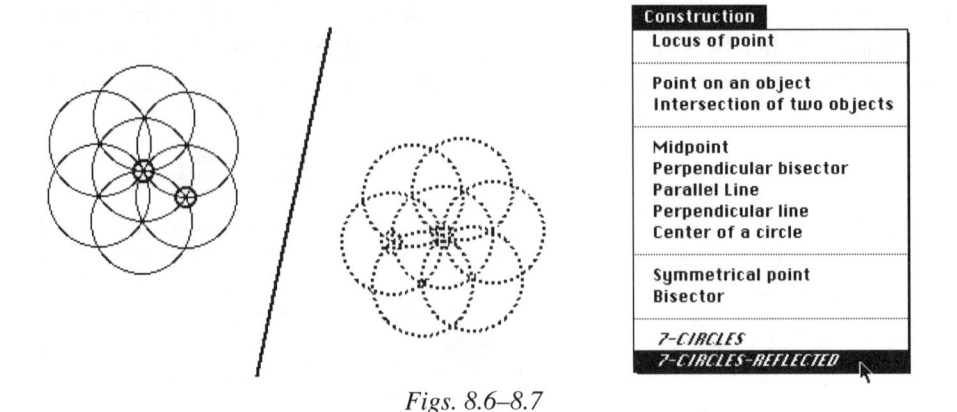

Figs. 8.6–8.7

Using the macros 7-CIRCLES and 7-CIRCLES-REFLECTED (Fig. 8.7), we produce Fig 8.8 (first reflecting in the horizontal line, then reflecting in the second line). When the lines are made orthogonal we obtain a figure with two axes of symmetry (Fig 8.9); this can be checked in the manner mentioned above.

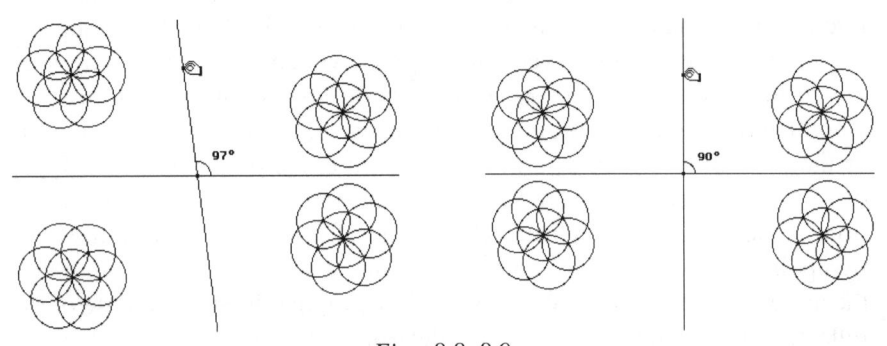

Figs. 8.8–8.9

This figure can be made to have even higher order symmetry. To do so, we connect the basic points of the initial figure to one of the angle bisectors of the two lines at right angles to each other, and we get a figure with four axes of symmetry (Fig. 8.10) which can be varied in an aesthetically pleasing way (Fig. 8.11, only one basic point need be varied). We have used the theorem (among others) that the successive reflection in n axes of a basic shape with n axes of symmetry produces a figure with n axes of symmetry provided one of these axes coincides with an axis of symmetry of the basic shape.

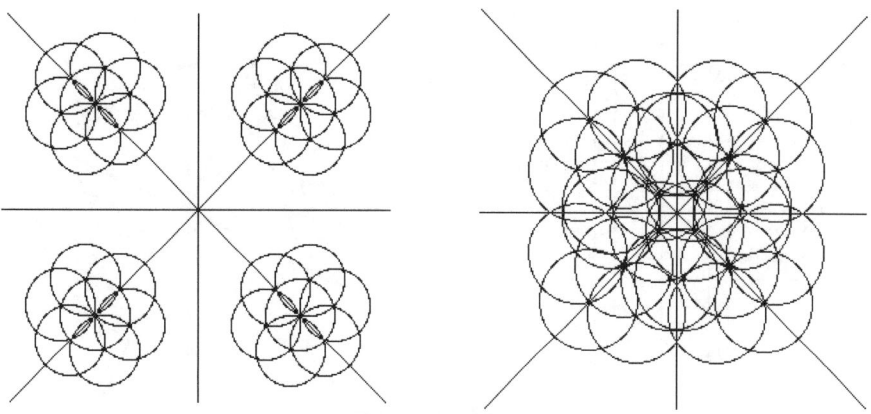

Figs. 8.10–8.11

Analogously, a figure with eight axes of symmetry may be produced (Fig. 8.12). Variations are shown in Figs. 8.13 and 8.14. By moving a basic point from an axis of symmetry the figure becomes one with just four axes of symmetry (Fig. 8.15, reduction of symmetry).

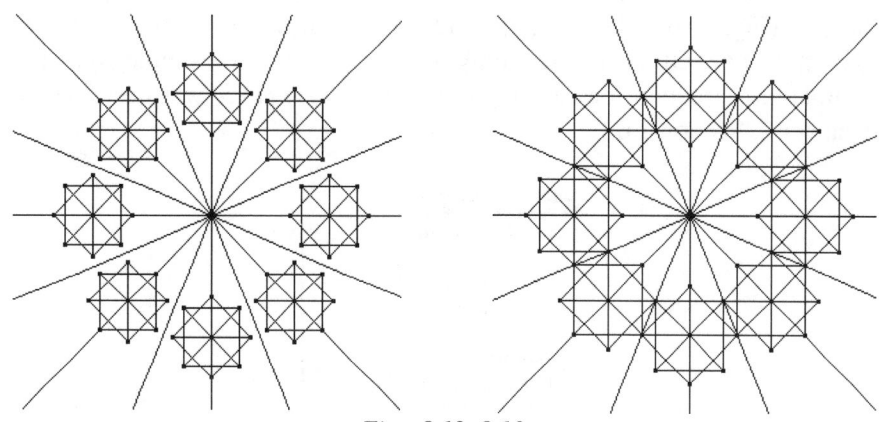

Figs. 8.12–8.13

Figs. 8.14–8.15

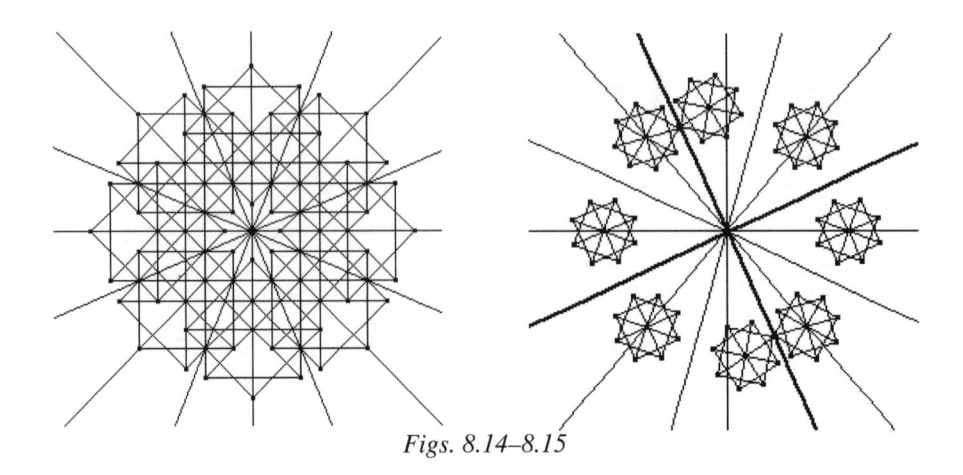

8.3 GENERATING LINE-SYMMETRIC FIGURES USING MACROS FOR REGULAR POLYGONS

Macros for the regular polygons with 3, 4, 5, 6, 8 sides, for example, may be used (Fig. 8.16). Macros are provided for all these on the disc associated with this book. Of those listed, only the pentagon is not a simple construction. (A construction for a pentagon *inscribed in a circle* was described in Chapter 2, Section 2.6.)

Construction
Locus of point
Point on object
Intersection
Midpoint
Perpendicular bisector
Parallel Line
Perpendicular line
Center of a circle
Symmetrical point
Bisector
EQUILATERAL TRIANGLE
SQUARE
REGULAR PENTAGON
REGULAR HEXAGON
REGULAR OCTAGON

Figs. 8.16

Generating geometrical figures with line symmetry

The initial objects each time are just two neighbouring vertices, from which the whole regular figure is constructed. By putting together and/or overlapping these regular building blocks, using corresponding macros, new figures develop, which can be less, equally or more symmetrical than the building blocks themselves. Fig. 8.17 has four axes of symmetry, Fig. 8.18 has six, Fig. 8.19 has eight. Of course, macros can be used in defining further macros, so that even more complex symmetric figures can be constructed.

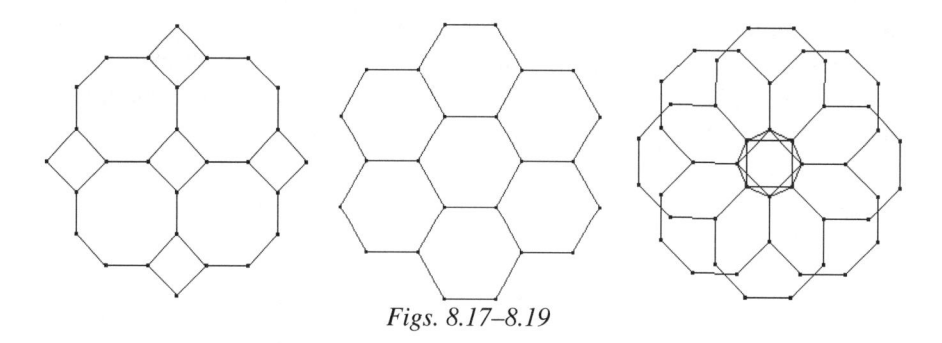

Figs. 8.17–8.19

8.4 GENERATING LINE-SYMMETRIC FIGURES USING VISUAL INSPECTION AND MEASUREMENT

8.4.1 Parallelogram

A parallelogram is constructed out of three points which determine it. We add the diagonals and centre lines to enhance the basic figure (Fig. 8.20). The initial parallelogram has no axes of symmetry. We vary a basic point to make the angles right-angles (Fig. 8.21) and get a rectangle with two axes of symmetry (passing through the midpoints of opposite sides). We pull the rectangle into a square (Fig. 8.22) with a further gain of two axes of symmetry (diagonals). We vary the square to produce a rhombus with only two axes of symmetry (diagonals) (Fig. 8.23).

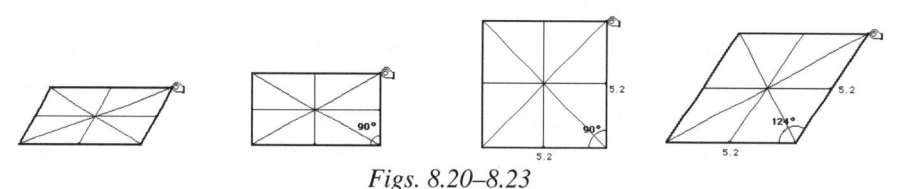

Figs. 8.20–8.23

147

8.4.2 Equilateral pentagon

With the help of the system macro **Circle by centre & rad. point** we can construct a freely jointed equilateral pentagon (Fig. 8.24, the help-lines – circles – are not shown). The five sides of the pentagon are copies of a free-standing line segment to the left of the figure. We connect a vertex of the pentagon to the perpendicular bisector of the opposite side using the option **Redefine an object** (or **Link a point to an object**). This forces the point to move and produces a pentagon with one axis of symmetry (Fig 8.25). Then we draw in the perpendicular bisectors of the remaining sides and vary the vertices (Fig. 8.26) until the perpendicular bisectors seem to meet at one point (Fig. 8.27). The interior angles of the pentagon are now equal (Fig. 8.28, regular pentagon with its five axes of symmetry).

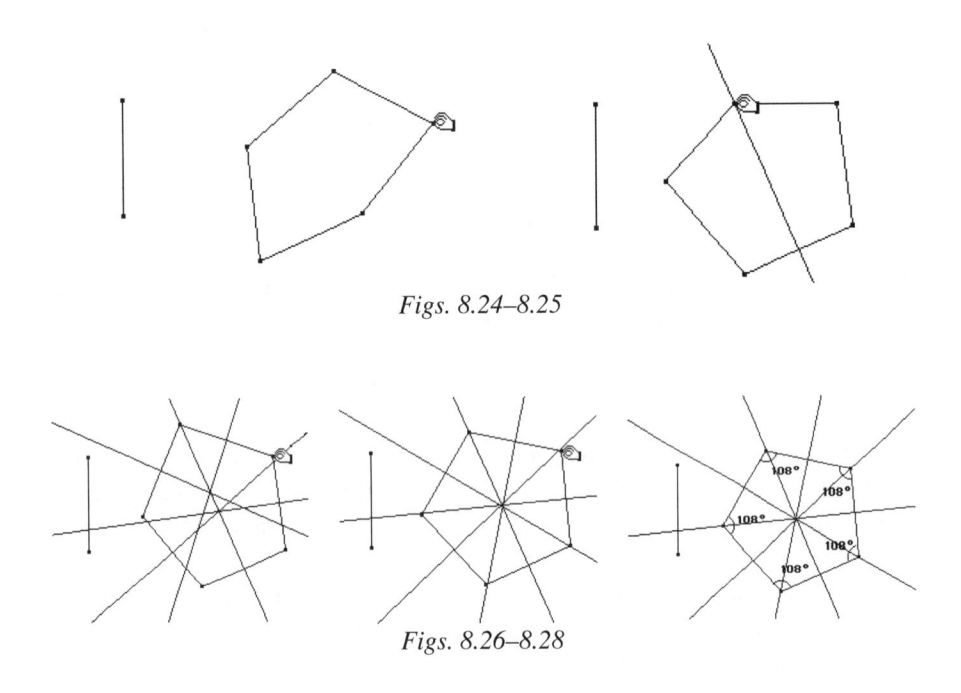

Figs. 8.24–8.25

Figs. 8.26–8.28

8.4.3 Affine-regular hexagon

Through the successive reflection of points of a base triangle in the perpendicular bisectors of its sides, we can produce an affine-regular hexagon (Fig. 8.29). We vary a vertex of the base triangle until its angles equal 60° and thus obtain a regular hexagon (Fig. 8.30, the axes of symmetry through the midpoints of opposite sides are not shown).

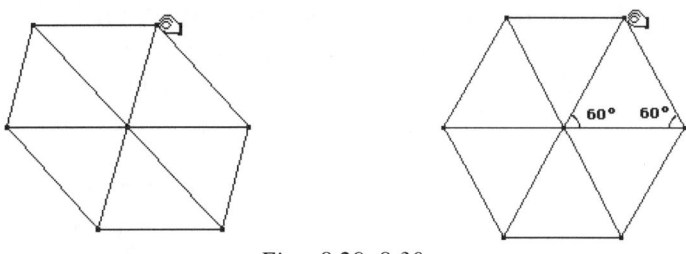

Generating geometrical figures with line symmetry

Figs. 8.29–8.30

8.5 GENERATING LINE-SYMMETRIC FIGURES USING FREE-HAND DRAWN LINES

The **Locus** command allows us to produce a line described by free-hand movement of a point; in addition, several points dependent on this freely moved point – reflectively symmetric points for example – can simultaneously produce loci of their own. Thus we can open the construction field for symmetric figures beyond those based on line segments, lines and circles to include those derived from totally free-hand figures.

If a point is reflected in a straight line (Fig. 8.31) and we move it so that it describes a continuous path, the reflected movements of the dragged point make the resultant figure symmetrical (Fig. 8.32). We choose the **Locus** command and, holding down the SHIFT key, click on the point to be dragged *and* on the reflected point, then the symmetric path is drawn at the same time as the path produced by the dragged point.

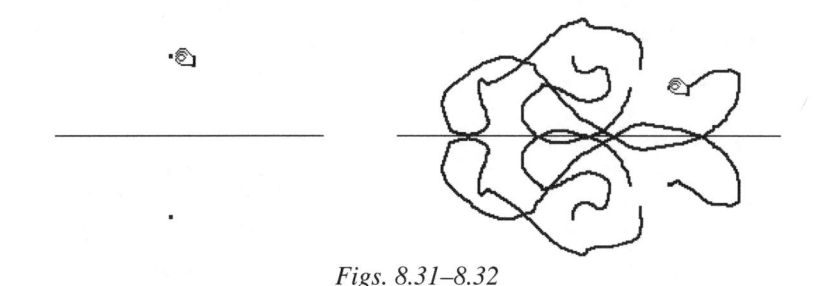

Figs. 8.31–8.32

We can also do this with a pair of points related by point-symmetry (Figs. 8.33–8.34). If we reflect a point (P) in *two* orthogonal straight lines in succession, the reflected points (P', P'', P''') display corresponding symmetric

149

Discovering Geometry with a Computer

movement when the basic point is moved (Fig. 8.35). With the **Locus** command we obtain a figure with two axes of symmetry (Fig. 8.36). (If we had only clicked on P" and P, a point-symmetric free-hand line would have appeared.)

Figs. 8.33–8.34

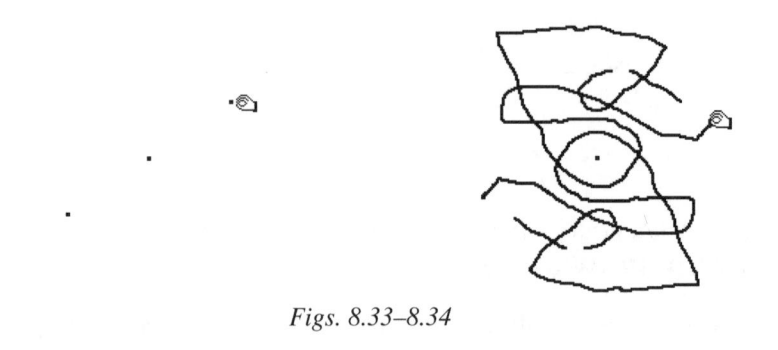

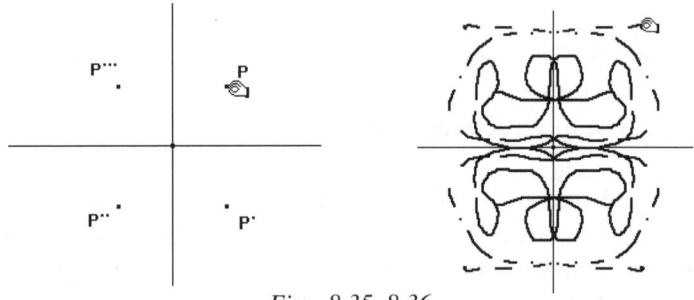

Figs. 8.35–8.36

In this way, we can produce n-way axis-symmetric free-hand patterns (Figs. 8.37–8.38 for $n = 3$; Figs. 8.39–8.40 for $n = 4$; Figs. 8.41–8.42 for $n = 5$). Highly symmetric frieze patterns can be generated in this way (Figs. 8.43–8.44).

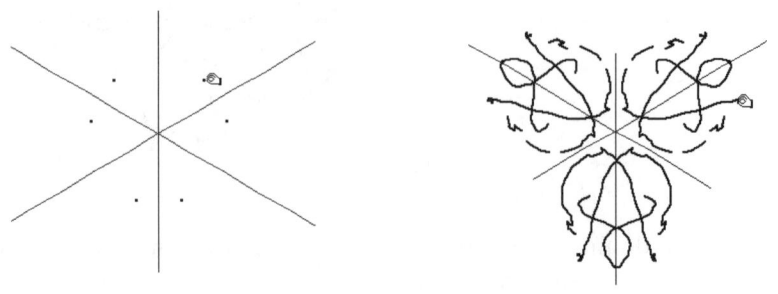

Figs. 8.37–8.38

Generating geometrical figures with line symmetry

Figs. 8.39–8.40

Figs. 8.41–8.42

Figs. 8.43–8.44

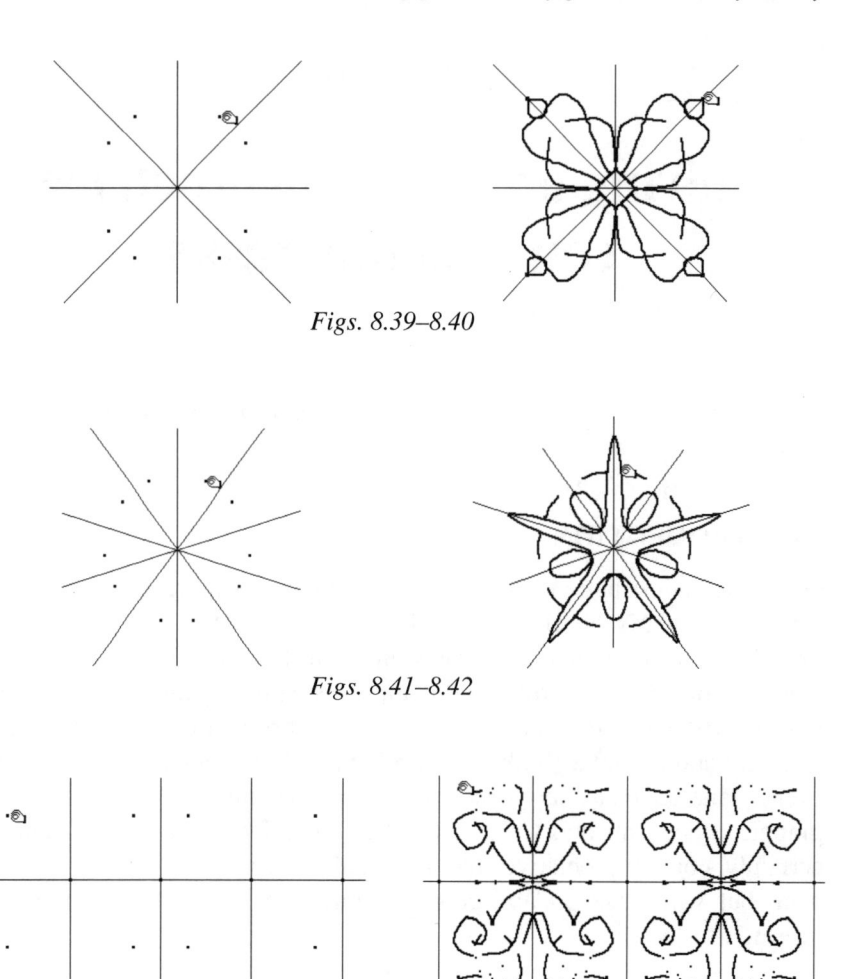

It is left to the interested reader to investigate further the 7 possible frieze patterns and 17 possible wallpaper patterns. (See also Chapter 11, Section 11.4.)

151

CHAPTER 9

Solving advanced geometrical construction tasks

Experimentation precedes construction.

9.1 INTRODUCTION

Geometric construction tasks can help students acquire problem-solving skills, particularly heuristic strategies. The computer can provide help in finding a construction solution as well as in the execution of the solution plan by simulating a ruler-and-compasses construction. The following examples demonstrate the efficiency of Cabri-géomètre for supporting heuristic and modular thinking in problem solving situations. In principle every plane geometric construction task can be solved with a suitable 2-D graphics system such as Cabri-géomètre. The selection of examples here is certainly not representative, the so-called insertion-tasks being predominant, but when using Cabri-géomètre many tasks can be interpreted as insertion-tasks.

9.2 EXAMPLES

Example 1 *Square in triangle – a new approach to an old problem*
We are given a triangle. We seek a square with two of its vertices on one side of the triangle and the other vertices one on each of the other two sides of the triangle. (This problem has been met before – in Chapter 6, Example 7 and in Chapter 7, Example 1; the approach here is slightly different again.)

First heuristic stage
We start with an acute-angled triangle ABC and construct two basic points on AB on which we draw a square with the macro SQUARE (Fig. 9.1).

Solving advanced geometrical construction tasks

Now we move a basic point until the corner of the square above the basic point comes to lie on AC (Fig. 9.2). We then do the same with the other basic point (Fig. 9.3); unfortunately the vertex of the square does not stay on line AC! After some further experimentation, we get the desired solution (Fig. 9.4). This ends the first heuristic stage.

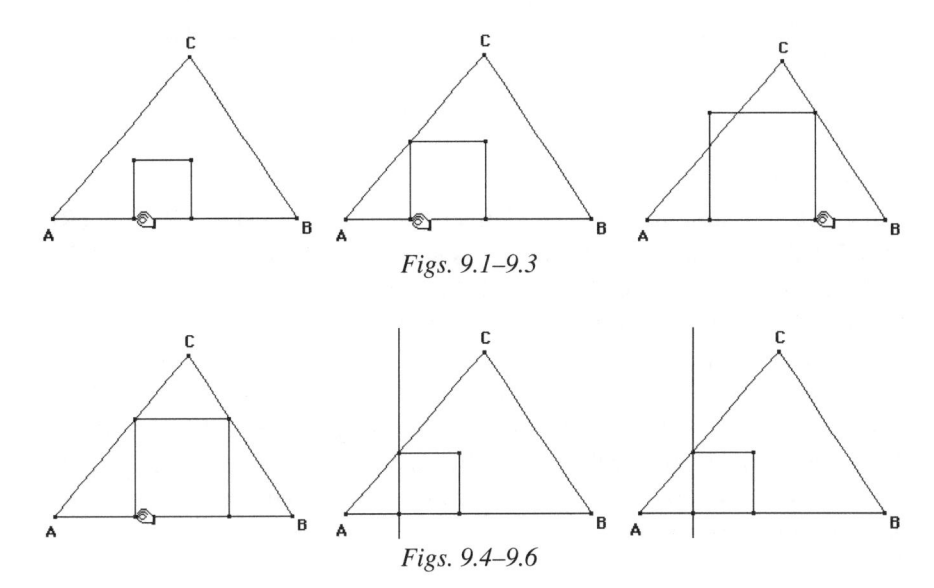

Figs. 9.1–9.3

Figs. 9.4–9.6

Second heuristic stage

It is easier to fulfil *two* of the given conditions at the same time. This is the second heuristic stage. We construct one basic point on AC and drop the perpendicular from it to AB (or we choose a basic point on AB and erect the perpendicular through it). We draw the square with one side delimited by the chosen point on AC and the foot of the perpendicular on AB (Fig. 9.5). We vary the chosen point sliding it along AC (Fig. 9.6) until the free corner of the square lies on BC (as in Fig. 9.4). In order to find a construction leading to the required point of intersection, we produce the locus described by the free corner of the square as one of the basic points is moved (Fig. 9.7; centre of enlargement A, the locus is a fixed straight line).

Thus we obtain the *idea* of a formal solution: construction of an auxiliary square (its vertices indicated by four dotted circles in Fig. 9.8), through whose free vertex and vertex A a line passes to intersect CB, followed by an enlargement about A of the auxiliary square to produce the desired square (result of using macro SQUARE shown in Fig. 9.8).

153

Discovering Geometry with a Computer

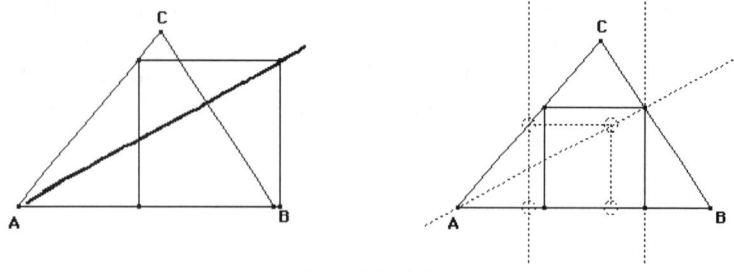

Figs. 9.7–9.8

Formal solution

Now this construction can be defined as a macro (SQUARE-IN-TRI-ANGLE). However, there is one proviso. We need a *specific* point on AC to start the construction, and the obvious choice is the midpoint. Thus we must start the construction again using the midpoint and then we will be able to save it as a workable macro. Having done that, we first mark the triangle *as an entity* as the initial object (bold in Fig. 9.9) and then mark the line segments of the solution square as the target objects (dotted in Fig. 9.9). Alternatively, we could choose the three vertices of the triangle as initial objects (bold circles in Fig. 9.10), which would lead to a more generally useful solution macro.

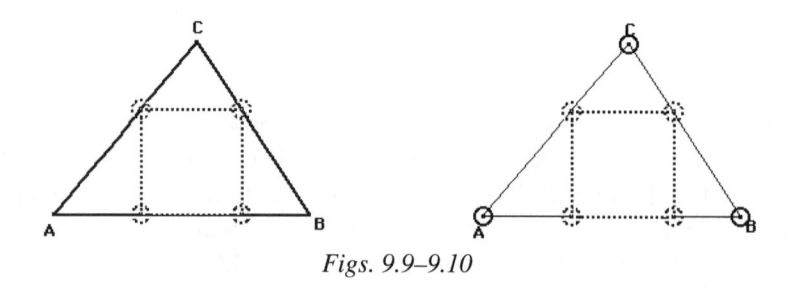

Figs. 9.9–9.10

As an experimental verification we produce an acute-angled triangle and use the solution macro: there is a solution for each side of the triangle (Fig. 9.11). We then change the triangle to a right-angled triangle, in which two solution squares coincide (Fig. 9.12). The obtuse-angled triangle has only one solution (Fig. 9.13)

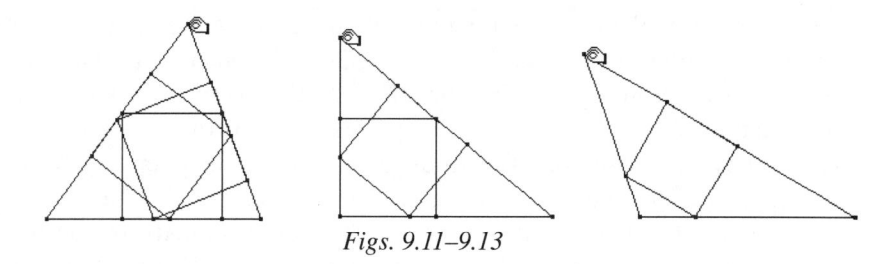

Solving advanced geometrical construction tasks

Figs. 9.11–9.13

Example 2 *Incircle of a triangle*

We are given a triangle. We seek a circle which touches the three sides of the triangle. We get an initial trial solution for a particular triangle by trying to fit a desired circle by inspection by varying the size of the circle radius and the position of its centre (Figs. 9.14–9.15). This circle loses its solution character immediately the triangle is varied, of course. We proceed systematically by making sure the touching radii are perpendicular to the sides. To do this, starting from a triangle` and arbitrary inside point, we construct the perpendicular line segments from the point to the sides of the triangle, then we draw a circle with this point as its centre and with one of the feet of the perpendicular as a circumference point (Fig. 9.16).

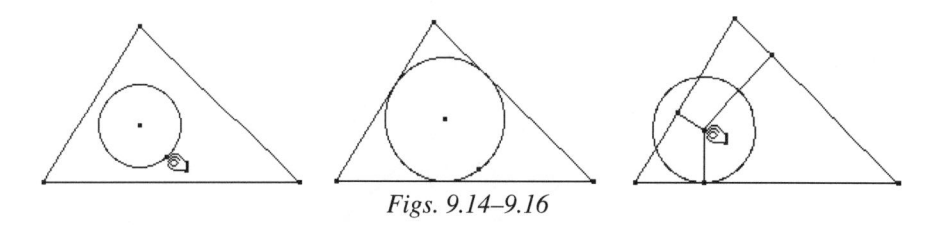

Figs. 9.14–9.16

Systematic variation of the centre provides the desired circle (Figs. 9.17–9.18), but, once again, variation of the triangle destroys the relationship.

Figs. 9.17–9.18

In order to reach a ruler-and-compasses construction which determines the circle, we make sure that the inside point lies on an angle bisector. We draw in a bisector, link the circle centre to it and move the centre until the circle touches the third triangle side (Figs. 9.19–9.20). The other two angle bisectors then also go through the incentre (Fig. 9.21); this must be so because the three touching radii are equally long. Thus the construction of the incircle is discovered. Variation of the triangle shows that the point of intersection property of the three angle bisectors is preserved and that this point of intersection is indeed the centre of the incircle (not illustrated). To construct the incircle, we need only two angle bisectors (Fig. 9.22). In defining the macro INCIRCLE, we can mark the whole triangle as a basic object (bold line segments in Fig. 9.23), or just the vertices as basic objects (small bold circles in Fig. 9.24) or the three individual sides of the triangle. The target objects are the incircle and its centre (shown dotted).

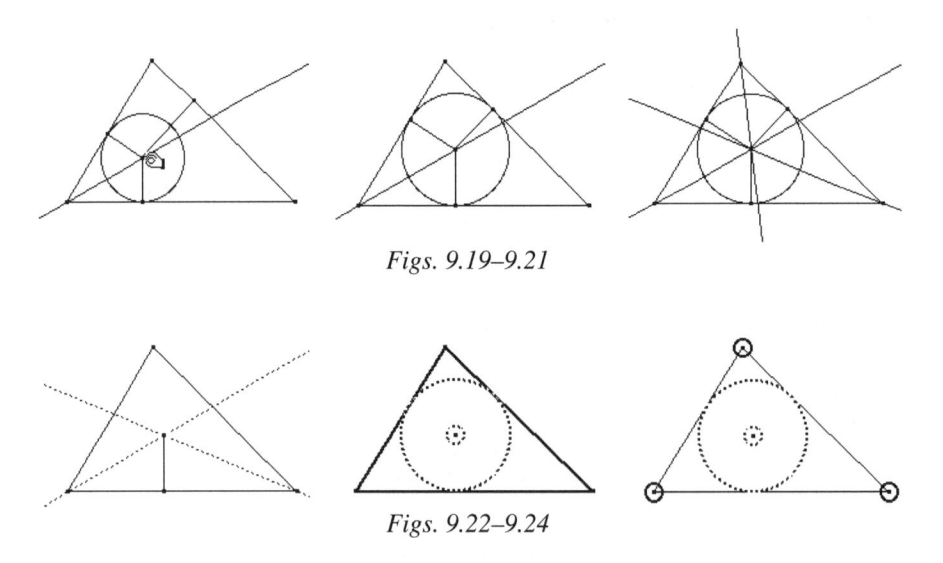

Figs. 9.19–9.21

Figs. 9.22–9.24

Example 3 *Trisected chord problem*
Given a circle and two radii, we wish to find the chord which is trisected by the two radii.

In Fig. 9.25 a particular initial configuration is given: a circle with midpoint M and radii MA and MB. A first experimental solution arises as follows: we construct two arbitrary circumference points X and Y, which serve as endpoints of the chord intersecting the radii (Fig. 9.26). We define the intersection points of the chord and radii and, using the appropriate **Miscellaneous** menu option, measure the three parts of the chord. Now we move the points X and Y on the circle until the length measurements are equal (Fig.

9.27). Thus an approximate solution is reached by measuring (with values to one decimal place), but it is not based on an algorithm or any clear procedure: the solution cannot be transferred to other initial data – it depends on the size of circle and positions of A and B. For each configuration a new trial is necessary.

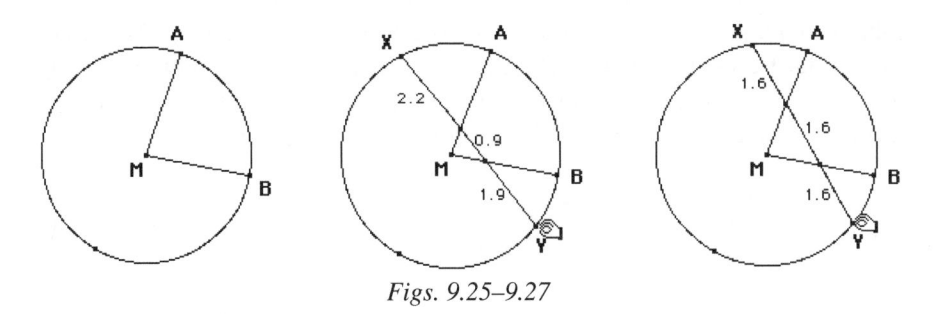

Figs. 9.25–9.27

Referring to Figure 9.27, we can see that XY must be parallel to an imaginary line segment AB (i.e. any perpendicular of XY is also a perpendicular of AB and the common perpendicular bisector of XY and AB passes through M). Therefore we draw the parallel to AB through an arbitrary circumference point Y, determine the relevant line sub-segments, whose lengths are measured (Fig. 9.28) and vary the position of Y until the chord is cut into thirds (Fig. 9.29). This method is more systematic – using the continuous shifting of the parallel line, by varying the position of just *one* point we can reach the solution. However, there is still no procedure available by which the desired chord can be constructed in a definite, *general* way.

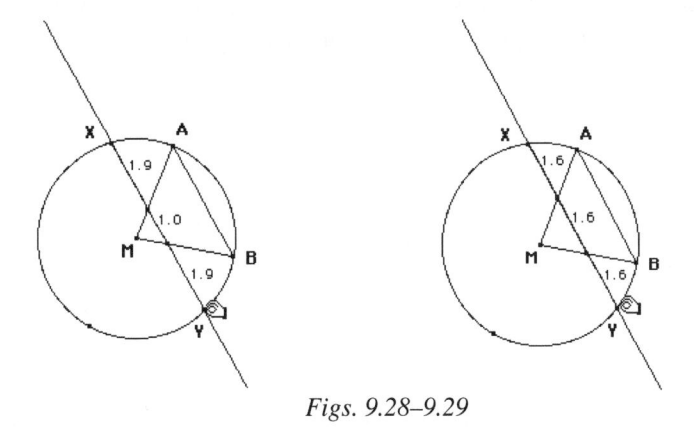

Figs. 9.28–9.29

Discovering Geometry with a Computer

So far, the experimental solution has depended on essentially interactive manipulation and dynamic measurement of objects. On the way to a general solution, we need to find appropriate heuristic strategies, like those described by Polya. Can we extend Fig. 9.29 by suitable auxiliary lines to help us? We draw the lines through M, X and through M, Y. These lines intersect the line through A, B, at X' and Y' respectively (Fig. 9.30). We have obtained a configuration of four rays emanating from M, giving three pairs of similar triangles. The trisecting of XY is transformed into the trisecting of X'Y'. (If we change the position of X or Y, the equal division is lost). We observe that |MX'| = 2|MX|. This ends the heuristic phase.

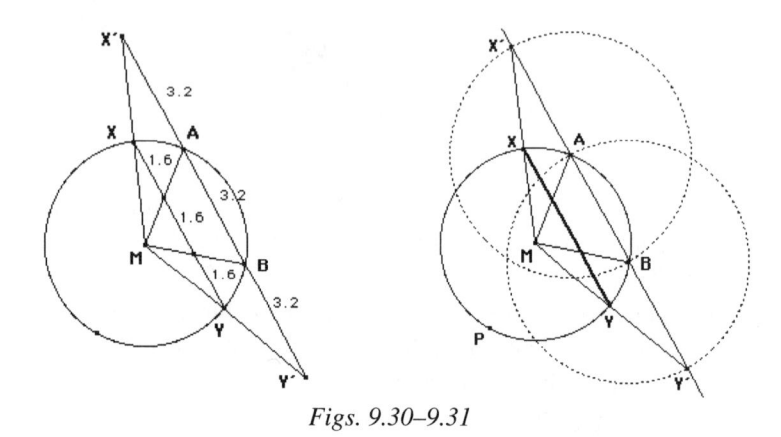

Figs. 9.30–9.31

Working backwards: now the general solution of the task can be achieved by simulating the corresponding ruler-and-compasses construction starting from the determining objects M, P, A, B (Fig. 9.31). All four determining objects can be varied; this is shown clearly if we vary the position of A (Figs. 9.32–9.34). Our construction is stable in drag-mode.

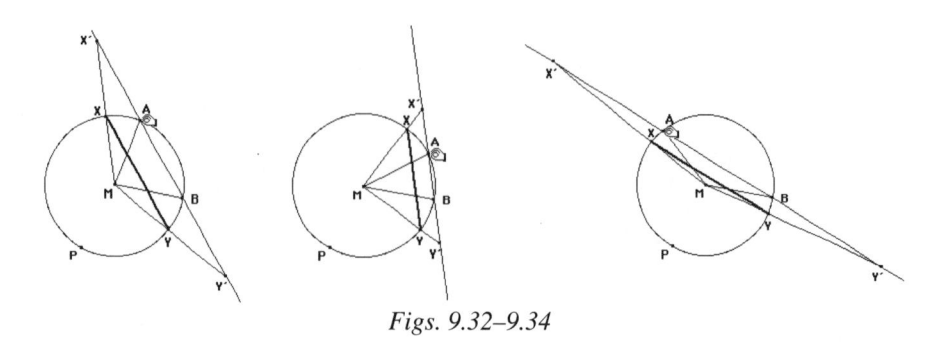

Figs. 9.32–9.34

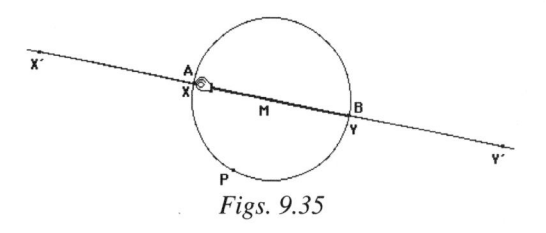

Solving advanced geometrical construction tasks

Figs. 9.35

Fig. 9.35 shows the degenerative case: A and B are endpoints of a diameter. The set task, therefore, has no solution for this case; in all other cases there is a unique solution, which is self-evident.

We are now in a position to define a macro for the general solution, by fixing the initial objects which determine the construction (indicated by four small bold circles in Fig. 9.36) and then marking the target objects which are to be constructed (dotted line segment and small circles representing points in Fig. 9.36). (The case of collinearity of A, B and M cannot be solved). We call the macro CHORD-TRISECTION (Fig. 9.37) and by clicking on the initial objects (i.e. circle centre, circle circumference point, chord end, chord end) in any appropriate initial configuration (Fig. 9.38) produce the desired chord with its points of division (Fig. 9.39).

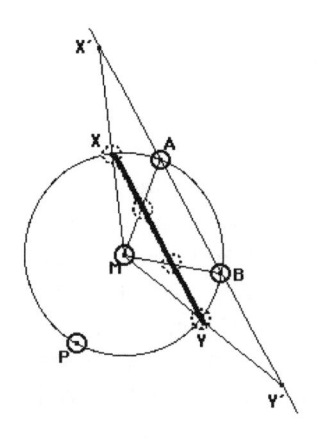

Figs. 9.36

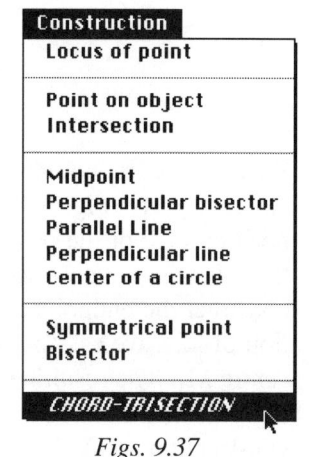

Figs. 9.37

Figs. 9.38–9.39

If we want to start from the given radii, the line segments MA and MB would have to be designated as the starting objects.

Example 4 *Midpoints of two circles*
We wish to construct a line segment given its midpoint and whose endpoints must lie on two given circles. Fig 9.40 shows an initial configuration. We choose a point X on the first circle centre M_1 and reflect X in P (Fig. 9.41) to produce Y.

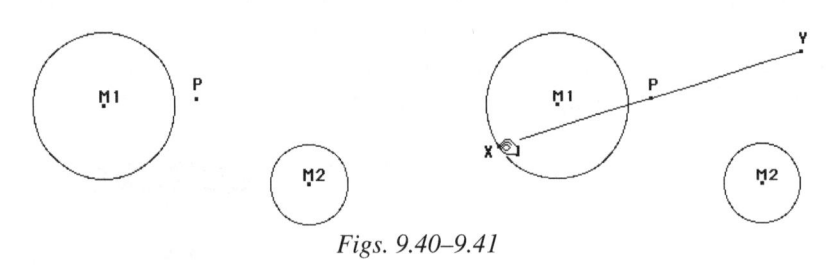

Figs. 9.40–9.41

We move X around the circle until its image point Y comes to lie on the circle centre M_2 (Fig. 9.42). If there is no solution possible, we change the position of P or the position or size of the second circle. In our configuration, we can obtain a second experimental solution (Fig. 9.43). In order to construct the endpoints of the solution line segments as points of intersection of loci, we ask for the geometric locus of Y as X moves on the circle centre M_1. The resulting locus is the image of the circle around M_1 reflected in P (shown bold in Fig. 9.44). Thus we obtain a ruler-and-compasses construction for Y_1X_1, Y_2X_2 to be carried out as a computer simulation (result in Fig. 9.45).

Figs. 9.42–9.45

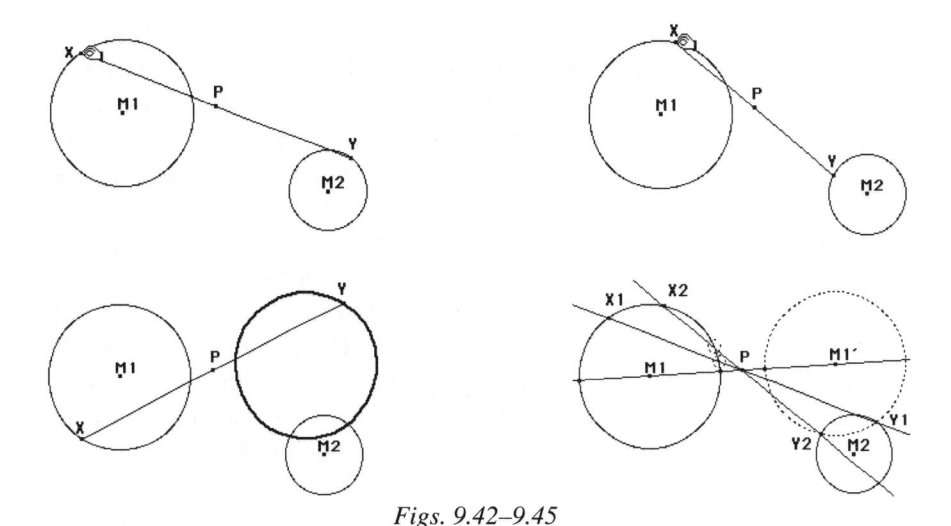

Now the corresponding macro definition can be obtained. The objects which determine the construction are: the first circle; the second circle; and P (each shown in bold in Fig. 9.46). The target objects are: Y_1, Y_2, X_1, X_2 and the desired connecting lines (each shown dotted; it suffices to click on Y_1, Y_2 and the two line segments). We name this macro CIRCLE-SEGMENT (Fig. 9.47). (Note: this macro could be defined slightly differently so that the initial objects were not the circles as entities but their centres and radius points.)

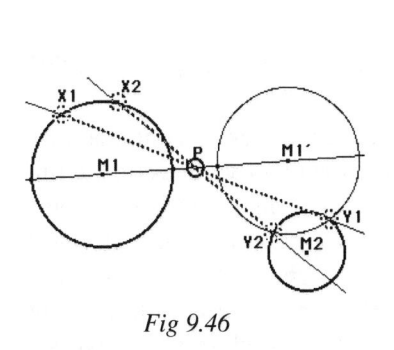

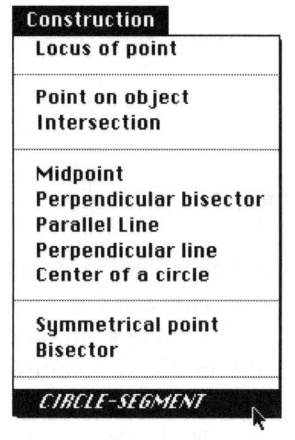

Fig 9.46

Fig. 9.47

We now consider a harder extension. The macro CIRCLE-SEGMENT is used on a particular initial configuration (Fig. 9.48). The target objects do not exist for this unusual configuration and the macro does not work. We investigate the solubility conditions for such a configuration. To do this, we vary the position of P until both solution line segments coincide (Fig. 9.49 shows the case where the two segments are about to become coincident). The geometric position of P must, in this case, be midway between the two given circles because P is defined as the midpoint of the line segment. The locus of this circle is the midpoint of the shortest line from the inner circle to the outer circle. This is the borderline case. Therefore we construct the circle whose centre lies on the line through the centres of the two given circles and which goes through the two midpoints of corresponding circumference points of the given circles i.e. where the two circles are closest and where they are furthest apart (Fig. 9.50). We link P to this circle and check our conjecture. If P lies outside the borderline circle, there is no solution; if P lies inside the borderline circle, there are two different solution line segments. Now we check again this finding for other positions of the given circles. Incidentally, for the borderline case of just one solution, in its extreme positions the line segment coincides with an outer common tangent of the given circles. Fig. 9.51 shows the situation as this limiting position is about to be reached.

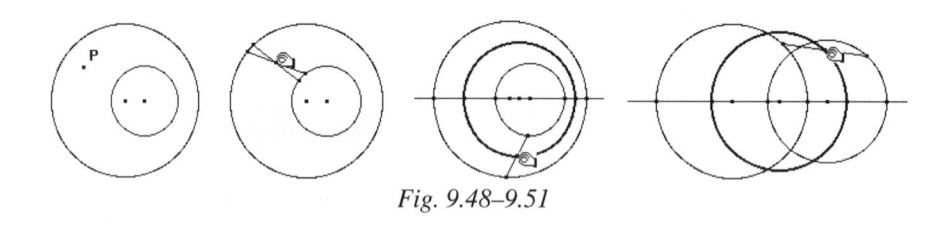

Fig. 9.48–9.51

Example 5 *Point, line, circle and equilateral triangle (PLC-EQTRI-ANGLE)*

We are given a point, a straight line and a circle. We seek an equilateral triangle with one vertex at the given point P and the other vertices on the line g and on the circle k. In order to reach an experimental solution, we omit the condition that one vertex lies on circle k. We choose a point on line g and produce a trial triangle with vertices on g and at P, using a macro EQUILATERAL-TRIANGLE (Fig. 9.52). By varying the position of the point on g, we find both solutions (Figs. 9.53–9.54); each solution involves a combined rotation and enlargement about P of the trial triangle. In order

Solving advanced geometrical construction tasks

to obtain a configuration providing the possible points of intersection, we ask for the geometric locus of the remaining free vertex of the triangle as the constrained vertex moves along g. We get a straight line (shown bold in Fig. 9.55). The point of intersection of this locus line with g coincides with a vertex of the equilateral triangle as one side of the triangle comes to lie on g (Fig. 9.56). The intersection line is parallel to the triangle side PQ (Fig. 9.57). We see that a 60° rotation of the triangle PQR about P to give image PQ'R' ensures that PQ' coincides with PR, and means that RR' must determine the required line.

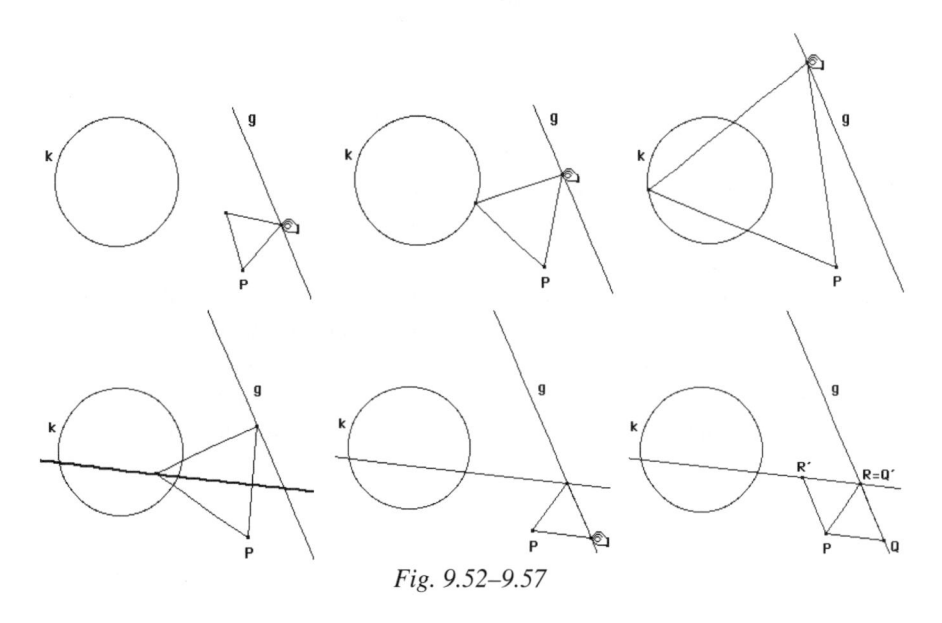

Fig. 9.52–9.57

Having completed the heuristic phase, a ruler-and-compasses construction can be simulated, the target being to find the desired triangle via the intersection of the circle and the image of g after a 60° rotation about P. The macro EQUILATERAL-TRIANGLE is used. The preferred method for the solution macro (PLC-EQTRIANGLE) is slightly different to that directly suggested by the above, but equivalent and easier to implement.

- A perpendicular is drawn from P to g meeting it at G (shown dotted in Fig. 9.58).
- An equilateral triangle PGH is erected (shown dotted in Fig. 9.58).
- A line is drawn through H perpendicular to PH (shown bold in Fig. 9.58). This is g', the required image of g, and intersects the circle (if a solution is possible).

163

- An equilateral triangle is drawn with one vertex P and another vertex a point where g' intersects the circle. This gives a solution triangle.

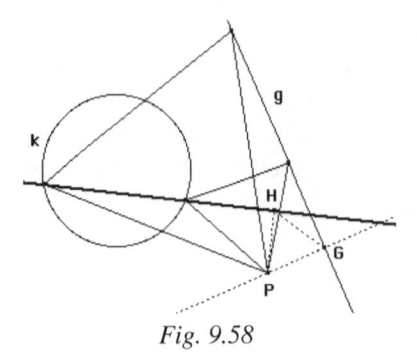

Fig. 9.58

In Fig. 9.59 the help lines have been hidden and P is about to be moved. If the image line of g is a tangent to the circle (Fig. 9.60) then there is exactly one solution triangle; if the image line misses the circle then no solution exists; in all other cases there are two different solutions. This result is independent of the position of the given objects, P, g, k, which can be checked from case to case. Using the construction result (Fig. 9.60), the construction sequence is now defined as a macro; the given objects are P, g and k (shown bold in Fig. 9.61 with target objects shown dotted).

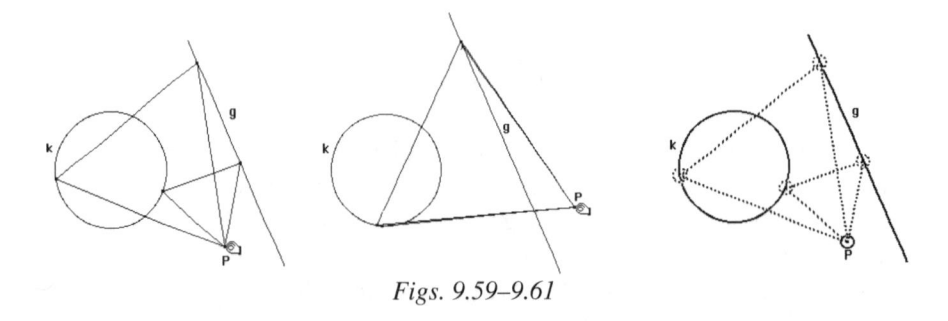

Figs. 9.59–9.61

Example 6 *Triangle constructed given a side, the opposite angle and the incircle radius (c, γ, r)*

We are given $c = 5$ units, $\gamma = 40°$ and $r = 1$ unit and are required to construct the triangle. We draw an arbitrary triangle but with c fixed as 5.0 units and construct its incircle possibly utilising a macro (see Example 2 for details of this construction). We also draw in a radius of the incircle to touch side c

(Fig. 9.62). We drag the γ-vertex of the triangle until we obtain the desired shape ($\gamma = 40°$) and size ($r = 1.0$) as in Fig. 9.63.

For a more systematic approach, we construct a triangle with side c = 5.0 units and connect the γ-vertex to the circumcircle of the triangle, whose centre is defined to be moveable on the perpendicular bisector of side *c*, and adjust the figure until $\gamma = 40°$ (Fig. 9.64).

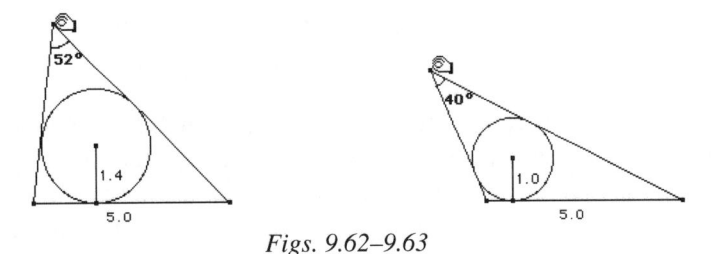

Figs. 9.62–9.63

We move the γ-vertex round the circumcircle, which keeps the circumference angle γ constant as the chord is fixed at 5.0 units, until the radius of the incircle becomes 1.0 (Fig. 9.65).

On our way to discovering a ruler-and-compasses construction of the solution, we investigate the locus described by the incentre (the centre of the incircle), as the γ-vertex moves around the circumcircle (Fig. 9.66, shown bold). We suspect it to be a circular arc.

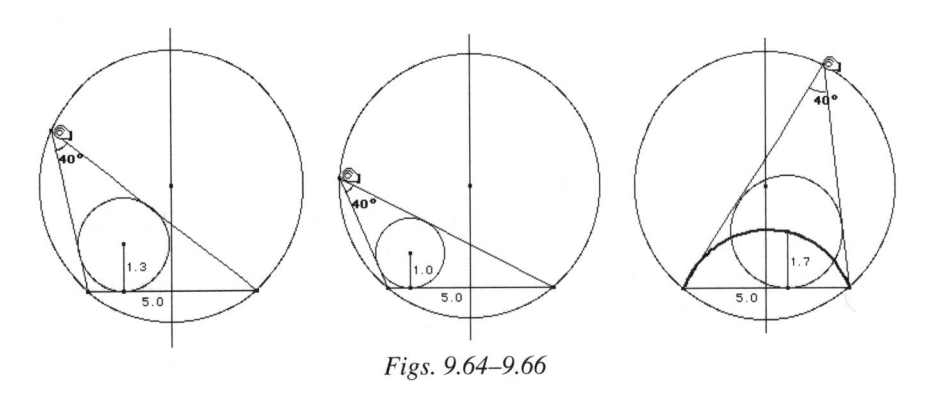

Figs. 9.64–9.66

If it *is* circular, then that arc must be part of a 3-point circle passing through the two endpoints of *c* and through the incentre, which can be checked experimentally. This circle is invariant during movement of the γ-vertex

165

Discovering Geometry with a Computer

round the circumcircle (Fig. 9.67). (It could be produced by the macro 3-POINT-CIRCLE). The centre of that 3-point circle is where the perpendicular bisector of *c* meets the circumcircle (we give no proof here). Thus we can construct one locus for the required incentre if we know the circumcircle (to be determined) and *c* (given); the other locus is a line parallel to *c* and *r* units from it (since the incircle radius *r* is given and the incircle must touch *c*. The two points of intersection of these two loci give the required positions of the incentres (Fig. 9.68 shows one solution, the other being symmetrical to it). Experimental checking of the solution is illustrated in Fig. 9.69.

Working backwards, we have to create tangents from the two endpoints of *c* to the incircle. Each tangent intersects the circumcircle at the γ-vertex which is to be found. A chord always subtends an angle at the centre twice that of any angle at the circumference. Thus the perpendicular bisector of chord *c* bisects the centre angle (2γ), see Fig. 9.69. Using these facts, we can construct the circumcircle starting with *c*. This ends the heuristic phase.

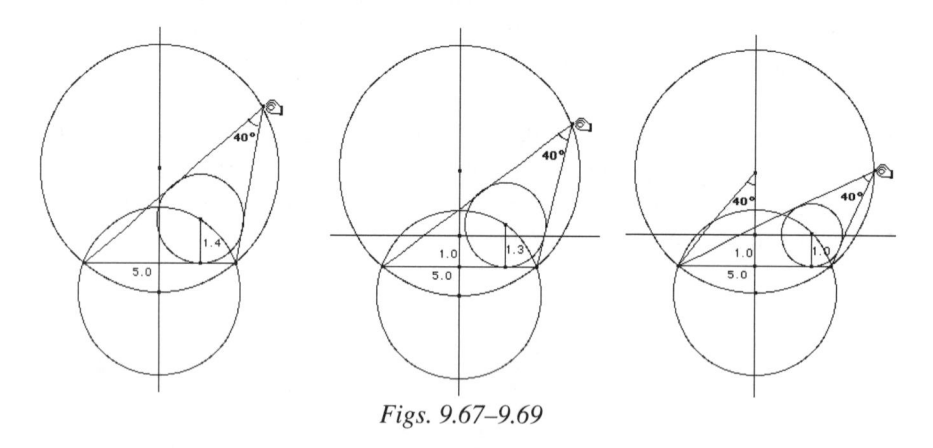

Figs. 9.67–9.69

We now take the three given values (*c*, γ, *r* to the left in Fig. 9.70) and a straight line and construct the desired triangle (Fig. 9.70, right) using the macros COPY-LENGTH, COPY-ANGLE, and TANGENT-PT-CIRCLE. The logical basis for the construction is as follows:

- The length *c* determines the base of the triangle which is placed on the supplied initial line.
- The angle γ and length *c* are used to locate the circumcentre which is the intersection of the perpendicular bisector of *c* and a line making an angle γ with it through an endpoint of *c*.

Solving advanced geometrical construction tasks

- The circumcircle can now be drawn.
- The perpendicular bisector of c and the circumcircle intersect at the centre of the 'circular locus' for the required incentre.
- The length r is used to construct a line r units from and parallel to c to provide the 'straight line locus' for the required incentre.
- The incentre is then found as the intersection of the two 'loci'.
- The incircle can now be drawn.
- A tangent to the incircle passing through one end of c is constructed and its intersection with the circumcircle found. This provides the third vertex of the required triangle, the other two vertices being the ends of c.

Unfortunately, this construction is too complex to be embodied in a macro for the currently available versions of Cabri-géomètre (PC 1.7, Macintosh 2.1). For such a macro Fig. 9.71 shows the given objects (small bold circles) and the target objects (dotted).

Despite not being able to construct a macro, we can vary the given values and observe the result. Fig. 9.72 shows r increased from that in Fig. 9.70, yielding a near-isosceles triangle (we guess that an exact isosceles triangle exists only as a limiting case; because of the coarse pixel structure we are not able to check this.). Fig. 9.73 shows r and γ changed, yielding another acute-angled triangle. Fig. 9.74 shows r and γ changed again, this time yielding an obtuse-angled triangle.

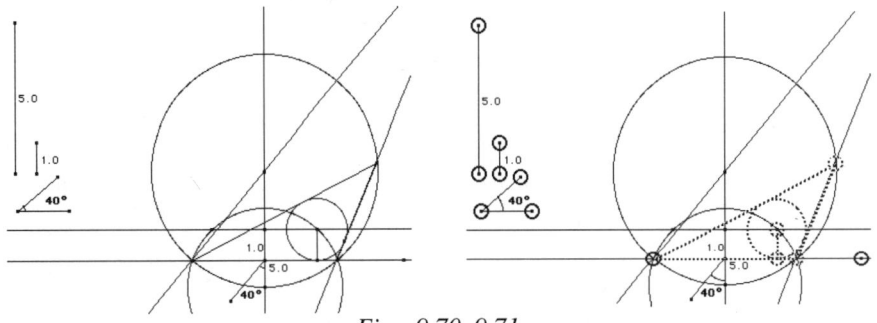

Figs. 9.70–9.71

Figs. 9.72–9.73

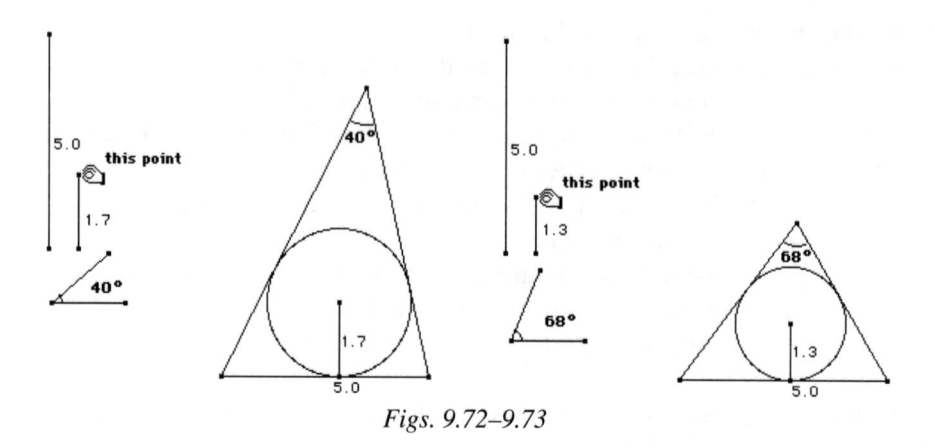

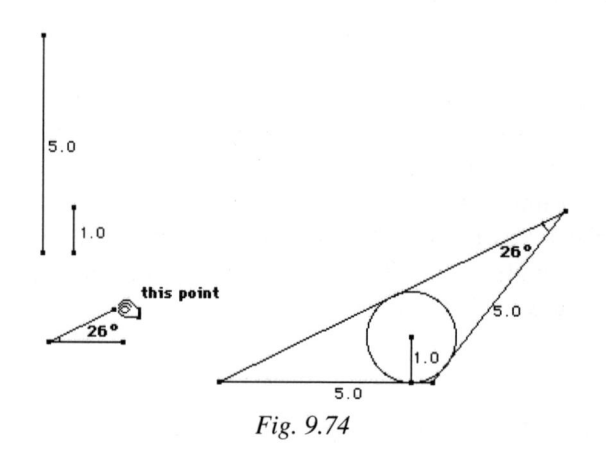

Fig. 9.74

CHAPTER 10

Exploring drag-mode geometry

Variety's the spice of life.
William Cowper

10.1 INTRODUCTION

An important feature of computer geometry systems such as Cabri-géomètre is drag-mode which represents an adaptation of the elastic band techniques of CAD systems and requires real-time computation necessitating a fast processor.

Drag-mode in an interactive 2D graphics system for construction in school geometry allows:

– Continuous variation on-screen of geometric diagrams (enhancing concept formation and theorem finding).
– Generation of loci (very difficult and time-consuming to do conventionally).
– Experimental discovery and checking of solutions to geometry construction tasks.

This opens up new possibilities of geometry learning and teaching based upon exploratory, individualised work. Basic geometric facts can be made more interesting and accessible. As geometric constructions become dynamic the gap is closed between the previously static construction results and the mentally conceived movements of geometric configurations often required of the learner.

This chapter begins with a review of the basic nature of drag mode, summarising ideas from previous chapters, and considers a number of elementary applications before moving on to advanced work in later sections.

Discovering Geometry with a Computer

10.2 DRAG-MODE AND ITS BASIC CHARACTERISTICS

In the following section, the dynamic sequence of events in drag-mode are documented using multi-phase 'snapshot' pictures (usually 2-phase).

10.2.1 Elementary geometric objects in drag-mode

The free mobility of geometric configurations in drag-mode requires the free mobility of those elementary geometric objects which constitute them.

- A point, P, is freely mobile (Figs. 10.1–10.2) and can be dragged to any position of the plane.
- A straight line (h), *implicitly* determined by point and direction, can be moved parallel to itself (Figs. 10.3–10.4).
- A circle, *implicitly* determined by its centre and radius, can be dragged without a change of radius (Figs. 10.5–10.6). (Note: to re-size hold down the OPTION key while dragging.)

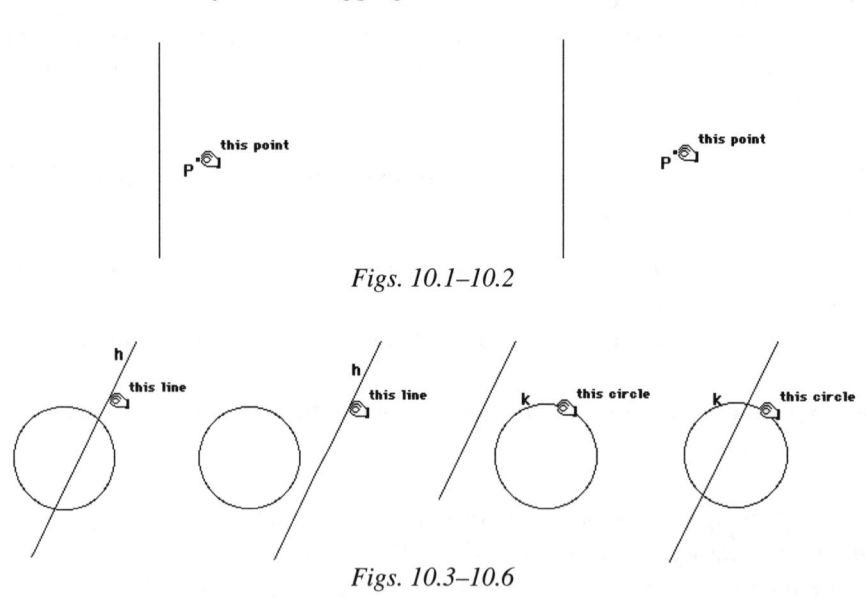

Figs. 10.1–10.2

Figs. 10.3–10.6

These are the dynamic properties of the three essential forms of basic object, and it is these properties which determine the possible variation of configurations in drag-mode.

The following elementary geometric objects depend on the points which define them.

Exploring drag-mode geometry

- The line through two points (P$_1$, P$_2$) can be rotated about a point (P$_1$) by dragging the other point (P$_2$) (Figs. 10.7–10.8).

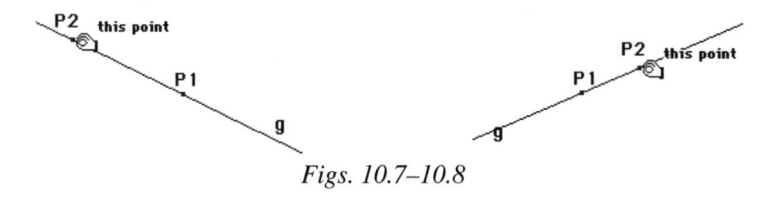

Figs. 10.7–10.8

- The same applies to the line segment connecting two points (Figs. 10.9–10.10). When line segments and angles with displayed measurements are changed by dragging an endpoint or vertex, the corresponding measurement values change simultaneously (Figs. 10.11–10.14).

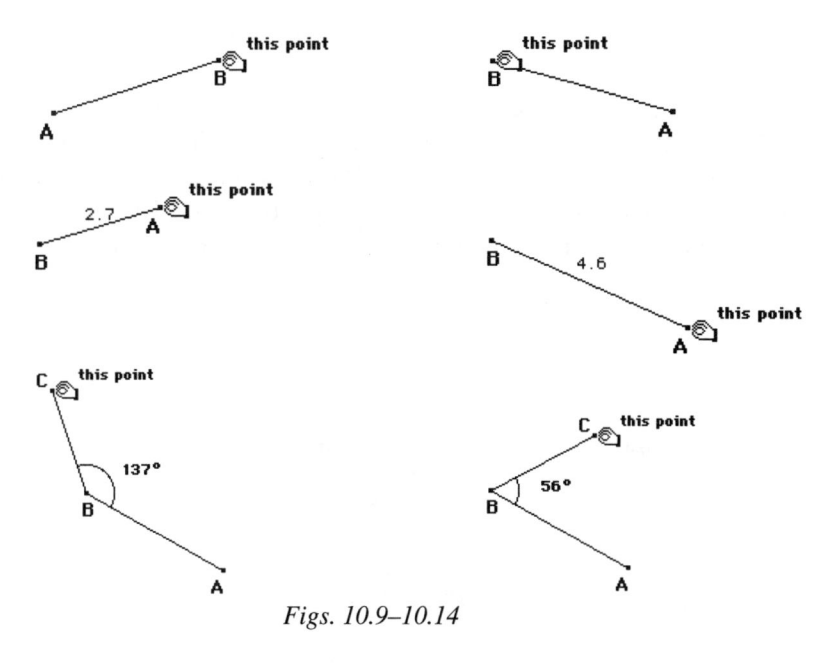

Figs. 10.9–10.14

- A circle, *explicitly* determined by centre (M) and circumference point (P), can be varied either by changing the position of the centre, in which case the circumference point remains fixed (Figs. 10.15 and 10.16), or by changing the position of the circumference point in which case the centre remains fixed (Figs. 10.17 and 10.18). It cannot be moved as an entity by dragging (unlike the implicitly defined circle).

Figs. 10.15–10.16

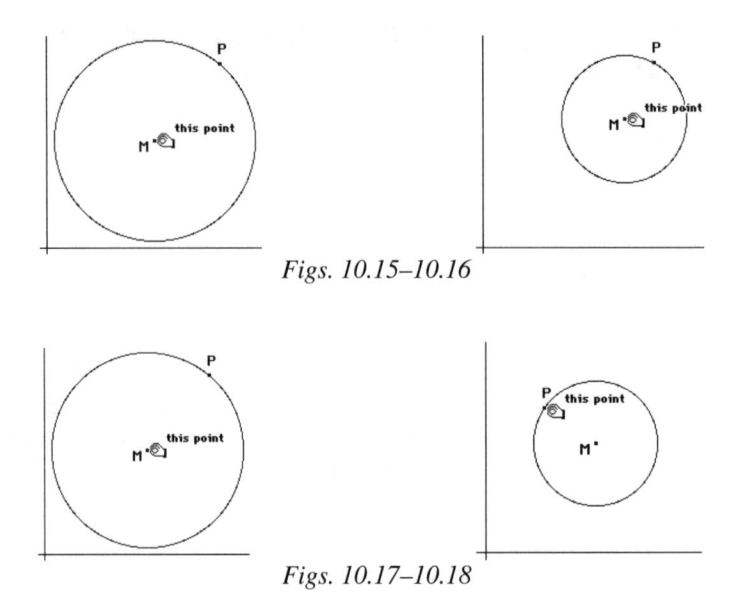

Figs. 10.17–10.18

A point (P) can be created on a straight line, line segment or circle (Figs. 10.19–10.21), or an already existing independent point can be connected to any of these objects using the **Miscellaneous** menu option **Redefine an object** (or **Link a point to an object**) (Fig. 10.22–10.24); it is then free to move along the line to which it is connected but cannot leave it.

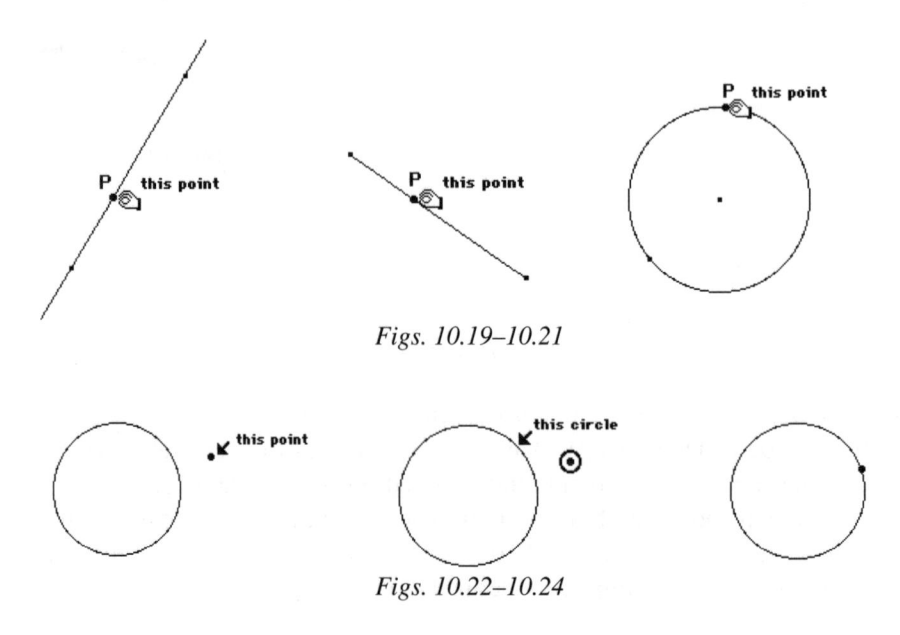

Figs. 10.19–10.21

Figs. 10.22–10.24

Exploring drag-mode geometry

In order to understand drag-mode the following working model may be helpful. Diagram 10.1 represents the following construction:

Starting from three basic points (A, B, C) the sides of a triangle are constructed (AB, AC, BC), then the altitudes (h_A, h_B, h_C) and finally the orthocentre (H).

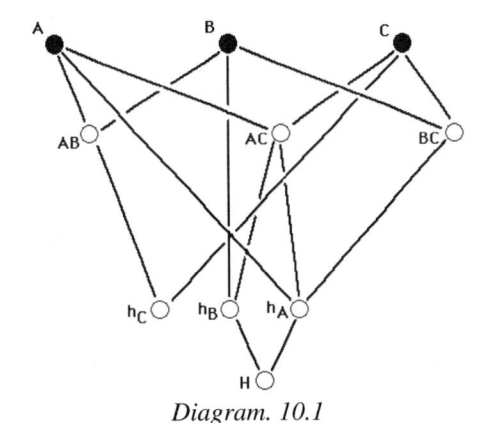

Diagram. 10.1

A construction starts from elementary objects, the so-called basic objects. All objects constructed from these basic objects are dependent upon them. Every completed part of a construction corresponds to a graphic function with the basic objects as input variables and the resulting objects (e.g. sides, altitudes, orthocentre) as output variables, which in turn can serve as inputs in further constructions. Every change of position of a basic object (carried out by the drag cursor) brings with it a change of position of all objects dependent on it. The screen updating occurs so quickly that the impression is given of continuous change of the configuration during the dragging of a basic object.

In the example which follows this process of 'continuous' change is documented using simple 2-phase pictures, first when vertex C is dragged (Figs. 10.25–10.26), and then when vertex A is dragged (Figs. 10.27–10.28).

Discovering Geometry with a Computer

Figs. 10.25–10.26

Figs. 10.27–10.28

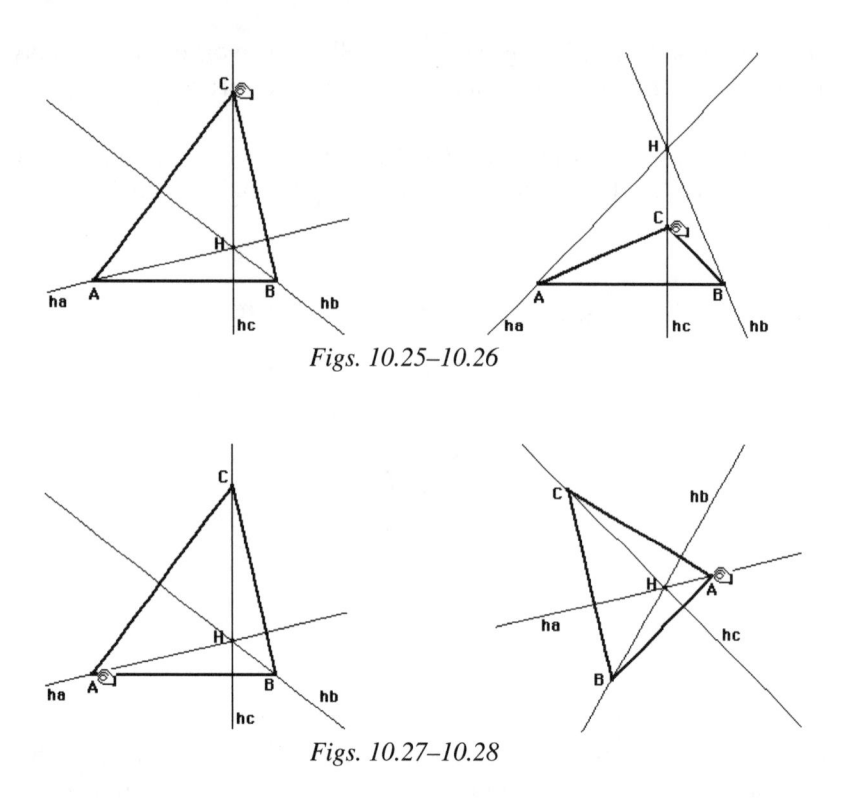

10.2.2 Invariant properties in drag-mode transformations

Transformations performed by drag-mode are straight line- and circle-invariant. The following relations (if constructively defined within figures and not purely fortuitous) are generally invariant during drag-mode transformations using Cabri-géomètre:

- parallelism
- orthogonality
- part-proportionality (i.e. ratio of lengths)
- point symmetry (rotational)
- line symmetry (reflective)
- incidence (in some cases).

We illustrate these characteristics by several examples.

- A trapezium (constructed to have a pair of sides parallel) is transformed into another trapezium by dragging a vertex (Figs. 10.29–10.30).

Figs. 10.29–10.30

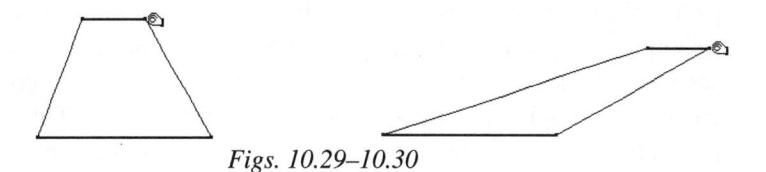

- The trisecting of the median of a triangle is preserved when dragging a vertex (Figs. 10.31–10.32).

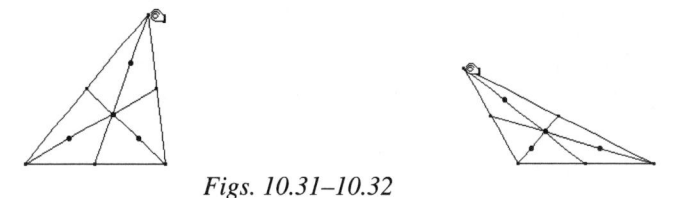

Figs. 10.31–10.32

- A point-symmetric polygon remains point-symmetric after the dragging of a determining vertex (Figs. 10.33–10.34).

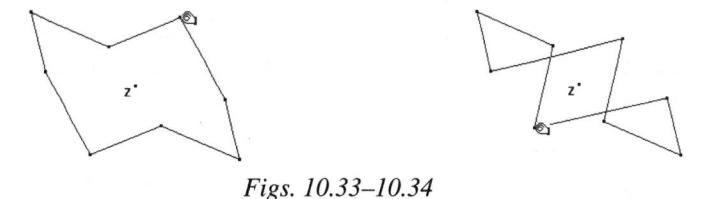

Figs. 10.33–10.34

- A line-symmetric polygon remains line-symmetric after the dragging of a determining vertex (Figs. 10.35–10.36).

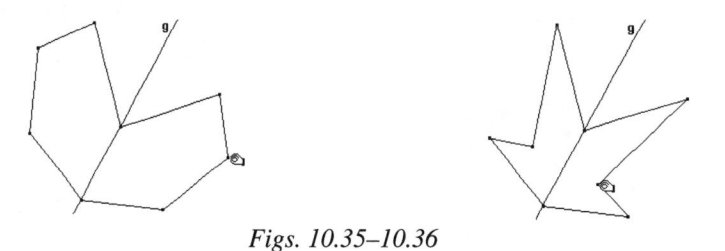

Figs. 10.35–10.36

- Concurrent points on objects remain concurrent in general when the objects are transformed in drag-mode. Points of intersection will disappear if the intersection is destroyed by dragging. In Fig. 10.37, a straight line intersects a circle; by dragging the straight line, the points of intersection coincide at a tangential point (Fig. 10.38), and then disappear when the straight line misses the circle (Fig. 10.39). This process is reversible provided the points of intersection have previously been defined using the **Construction** menu option **Intersection (of two objects)**.

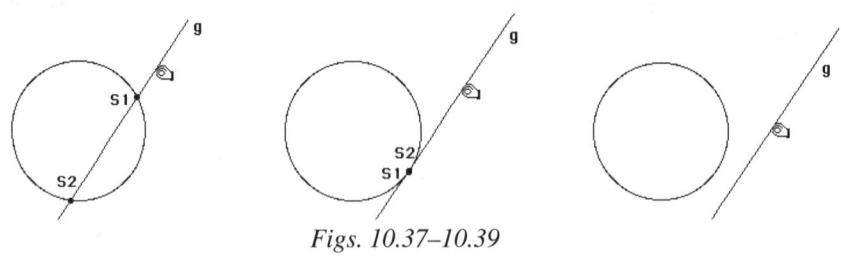

Figs. 10.37–10.39

- A triangle is created and the perpendicular bisectors of its sides are constructed and their common point of intersection defines the circumcentre and the circumcircle drawn (Fig. 10.40). The triangle is then changed by dragging one vertex (Figs. 10.41–10.42); the perpendicular bisectors remain perpendicular bisectors, the circumcircle remains a circumcircle. When the triangle is degenerated the perpendicular bisectors are parallel and thus have no point of intersection, and there is no circumcircle (Fig. 10.43).

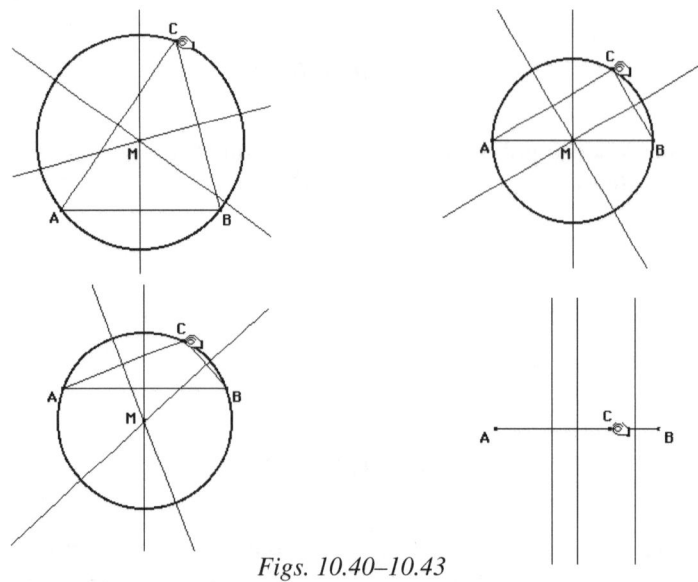

Figs. 10.40–10.43

Exploring drag-mode geometry

- The orientation of a figure or a line may or may not be consistent after drag-mode (i.e. the figure may or may not be 'turned over').

  (a) On a line through P and Q is drawn the side AB of an equilateral triangle ABC where A and B are basic points and C is the dependent point (Fig. 10.44). If we drag P towards Q (or Q towards P), we observe that the orientation of the triangle is preserved (A,B,C form an *anticlockwise* triplet) even when P and Q pass each other and the triangle inverts (Fig. 10.45).

  (b) If we drag A past B the triangle inverts, but once again the orientation of the triangle is preserved.

  (c) If a *basic* triangle ABC is created as in Fig. 10.44 but with C an independent point, then if C is dragged across the line through AB then the orientation is *not* preserved and A,B,C now form a *clockwise* triplet (not illustrated).

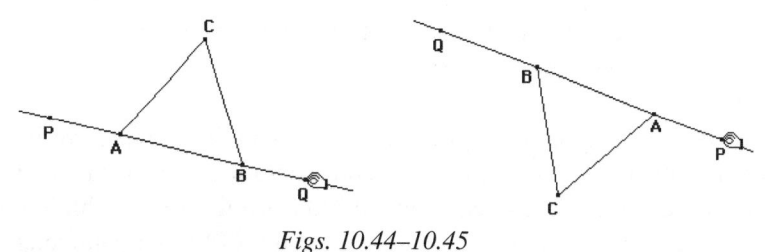

  Figs. 10.44–10.45

  (d) If a figure is constructed using **Symmetrical point**, or by other means to reflect a constituent point in a line, and such a point is dragged across the line of symmetry the orientation of the figure is *not* preserved. Figs. 10.46–10.47 illustrate this – P' is the image of P under reflection in the line through A, B.

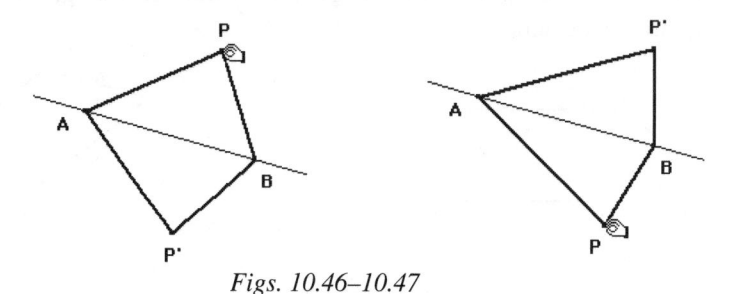

  Figs. 10.46–10.47

10.3 ELEMENTARY APPLICATIONS OF DRAG-MODE

The uses of drag-mode represented in the following were selected primarily from the point of view of exploring geometric facts. The selection is constrained by limitations of space and the practical uses of synthetic basic geometry are not considered (see Chapter 11).

10.3.1 Polygonal figures in drag-mode

The treatment of polygonal figures in drag-mode has many didactic consequences. It is mainly important for:

- allowing a large number of figures to be studied to illustrate various basic polygonal concepts (e.g. What is a triangle?);
- illustrating theorems relevant to school geometry relating to polygonal figures;
- promoting the study of transformations and operations (represented by movement of polygonal figures).

There are many possible ways to vary figures in drag-mode. We illustrate this using examples concerning a parallelogram and a hexagon. In Fig. 10.48 a parallelogram has been constructed starting from one basic point A lying on the line through A, E, and using copies of three parameters (one angle and two sides) which appear in Fig. 10.48 beneath the parallelogram itself. (How to copy angles and lengths was discussed in Chapter 3; macros for these are supplied on the disc associated with this book.) Only A of the four vertices A, B, C, D can be freely moved in drag-mode, the other three being dependent. The parallelogram can be rotated about A (or E) if AB lies on the line through A, E (Fig. 10.49 shows rotation about A by dragging E). In this way a same-orientated congruent positional change of the parallelogram can be achieved.

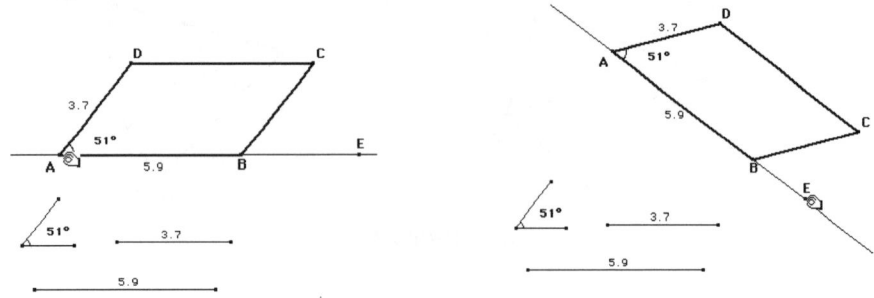

Figs. 10.48–10.49

Exploring drag-mode geometry

If we eliminate one of the determining line segments as a parameter, a further point becomes a drag point (its distance from A now being freely chosen); in our example this is D (Fig. 10.50). By dragging vertices D or A, AD can be lengthened or shortened at will, but AB is of fixed length as it is a copy of the determining line segment situated below the parallelogram, and, similarly, angle DAB is fixed (Fig. 10.51; the action is an axial affinity).

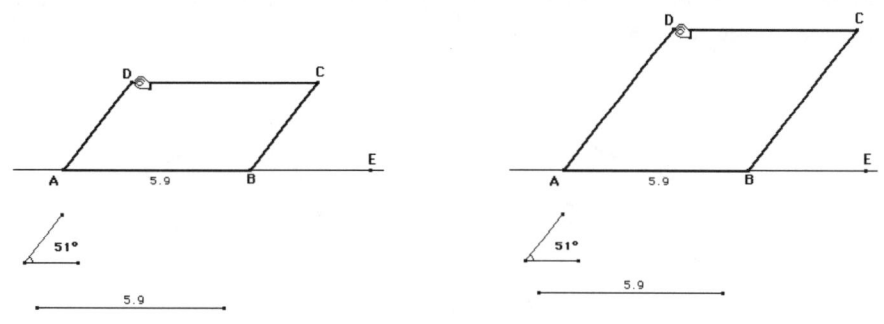

Figs. 10.50–10.51

If we now eliminate the second line segment as a parameter and take the angle to be the sole determining object, we gain B as a further drag point (along with A and D); under dragging we obtain nothing but parallelograms all with the same angles (not illustrated).

If we allow just the two line segments to be the determining factors, and not the angle, we obtain a flexible linked parallelogram (Fig. 10.52), whose vertex D can be moved on a circle, centre A. Fig. 10.53 shows AD being rotated by dragging a point placed on it, causing D to follow the circular path shown as an interactively generated locus. (Section 10.4 provides further examples of locus generation).

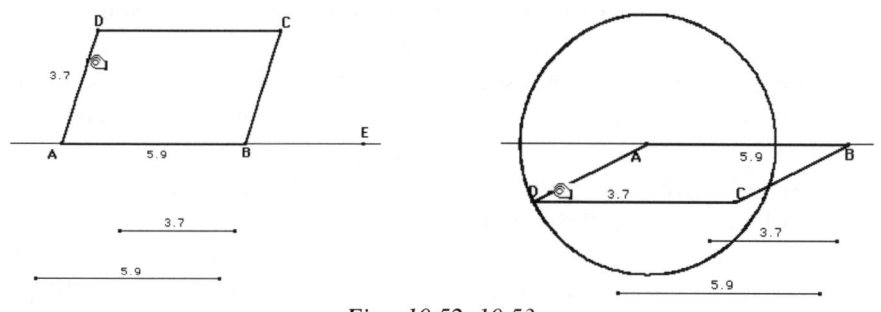

Figs. 10.52–10.53

The construction of a parallelogram without the limitation of *any* determining parameters allows the parallelogram to be freely varied by dragging the three corners which determine it (here A, B, D). Thus with drag-mode we can generate all parallelograms, as far as the (virtual) screen allows (Figs. 10.54–10.55 show variations by dragging B). We could also achieve that by varying the three determining parameters (shown earlier in Fig. 10.48). For example, Figs. 10.56–10.57 illustrate changing the angle and both line segments by dragging. Note that Fig. 10.56 can be considered as representing a function with three input variables.

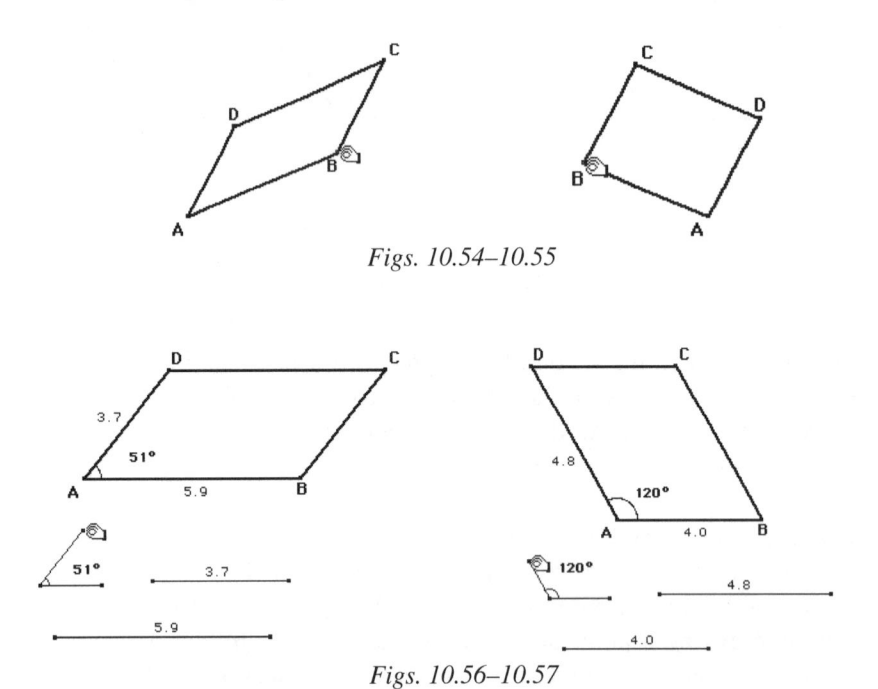

Figs. 10.54–10.55

Figs. 10.56–10.57

In Fig. 10.58 we construct a regular hexagon on AB (which could also be produced with a macro). A and B are the points which determine the hexagon. We then move B and thus generate a rotation-enlargement of the hexagon with A as the centre (Fig. 10.59). Vertex C is of course dependent on A and B. However, it can be disconnected from the construction of the hexagon using the **Miscellaneous** menu option **Redefine an object** (or **Delete relation**) and selecting **Basic point** (N.B. *not* **Point on object**!). C can then be dragged away as an independent point. A 'locally axis-affine' variation of the hexagon results. Only C moves out of the regular hexagonal shape (Figs. 10.60–10.61).

180

Exploring drag-mode geometry

Figs. 10.58–10.61

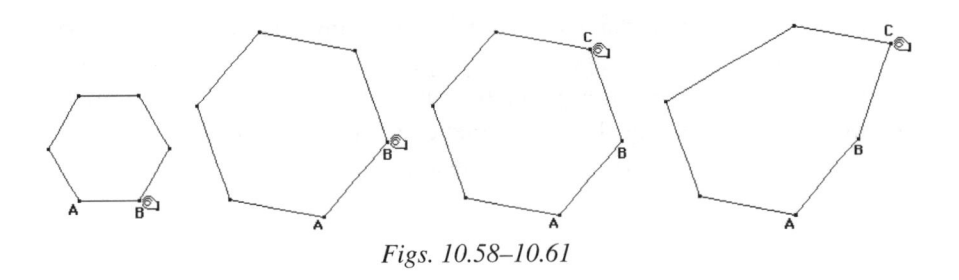

Drag-mode can be used on polygons as:

- a same-orientation congruent transformation (Figs. 10.46–10.47);
- a same-orientation similarity transformation (Figs. 10.58–10.59);
- a (locally) axis-affine transformation (Figs. 10.60–10.61).

Over and above that, a polygon can be varied as a freely-jointed linkage in various ways if its sides are basic objects. Thus, a flexible polygon is varied by dragging a vertex in Figs. 10.62–10.63; the four determining lengths are shown to the left of the figure, and are copied to produce the polygon itself (see Chapter 3 for information on copying).

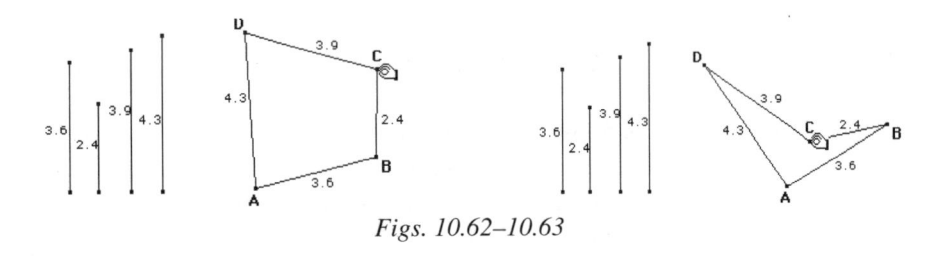

Figs. 10.62–10.63

10.3.2 Drag-mode in theorem finding

From the wealth of possible examples we choose some to show the efficiency of drag-mode in discovery-learning in elementary geometry. The principle of invariance is fundamental to theorem finding: which properties of a configuration remain invariant (and which change) under drag-mode transformations?

Example 1 *Quadrilateral formed by the perpendicular bisectors of the sides of a parallelogram*
The perpendicular bisectors of the sides of a parallelogram form another parallelogram (shown bold in Fig. 10.64), whose position and shape varies

according to the dimensions of the initial parallelogram (Figs. 10.65–10.67). How does the shape of the perpendicular bisector parallelogram depend on that of the initial parallelogram? We 'specialise' the parallelogram to a rectangle (Fig. 10.68: a degenerate case, producing a point) or a rhombus (Figs. 10.69–10.70: producing a rhombus).

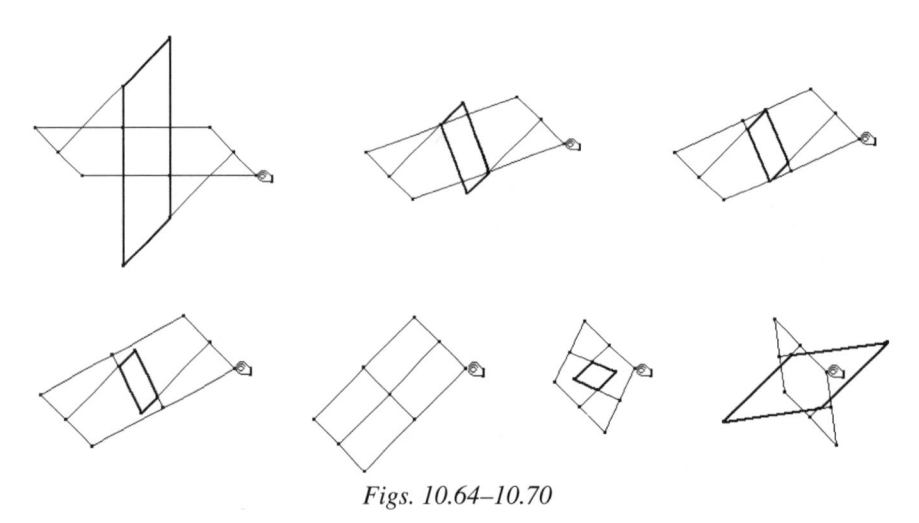

Figs. 10.64–10.70

Example 2 *Cyclic quadrilateral (from German National School Mathematics Competition 1989)*
A cyclic quadrilateral is drawn (Fig. 10.71) and rectangles are drawn on its four sides, such that in each case the length chosen for the second side of the rectangle is equal to that of the opposite side of the cyclic quadrilateral (Fig. 10.72). The four centres of these four rectangles are joined to form a quadrilateral. What shape is it?

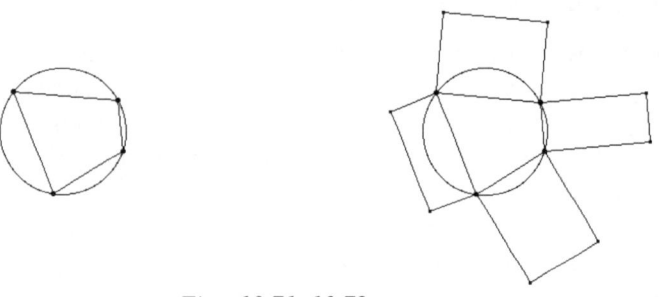

Figs. 10.71–10.72

Exploring drag-mode geometry

Figs. 10.73–10.77 show a varying sequence; we see that we always get a rectangle (we call it the mid-rectangle). For which cyclic quadrilateral does the mid-rectangle become a square? In Fig. 10.77 the cyclic quadrilateral is degenerated to a triangle and the mid-rectangle still exists!

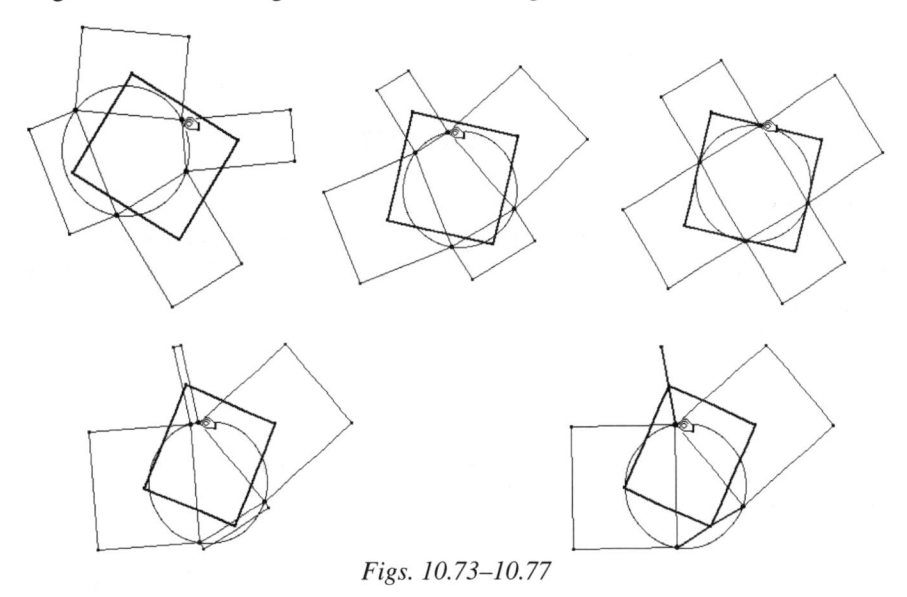

Figs. 10.73–10.77

We modify Fig. 10.77 to produce the limiting configuration when the circle is reduced to a single point (Fig. 10.78). What conjecture arises? We leave it to the interested reader to formulate and disprove or substantiate a conjecture.

We obtain a pleasing figure for the special case of the equilateral triangle (Fig. 10.79).

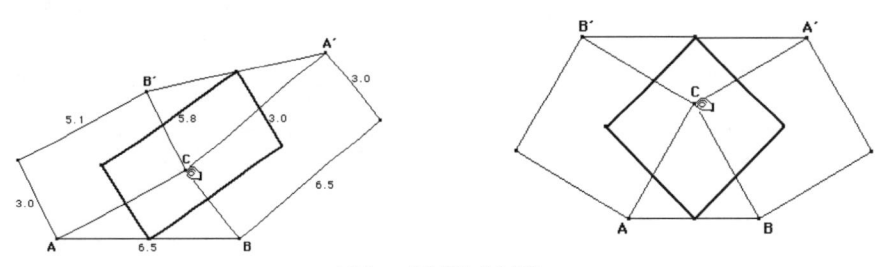

Figs. 10.78–10.79

Example 3 *Circle-tangent quadrilateral*
Hint: The sum of lengths of certain pairs of sides … (Figs. 10.80–10.81).

183

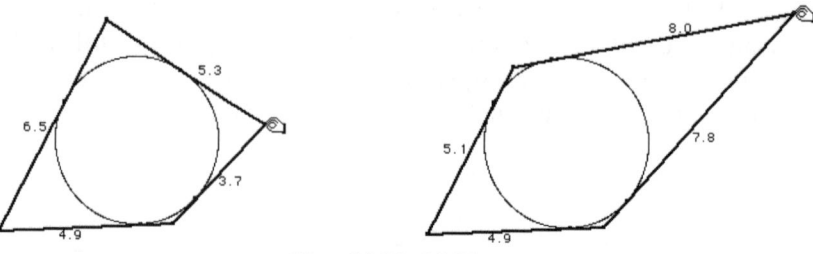

Figs. 10.80–10.81

Example 4 *Circle-tangent hexagon*
In a circle-tangent hexagon, the line segments connecting each pair of opposite vertices intersect at one point (Figs. 10.82–10.83) – Brianchon's Theorem for the circle as a particular conic section.

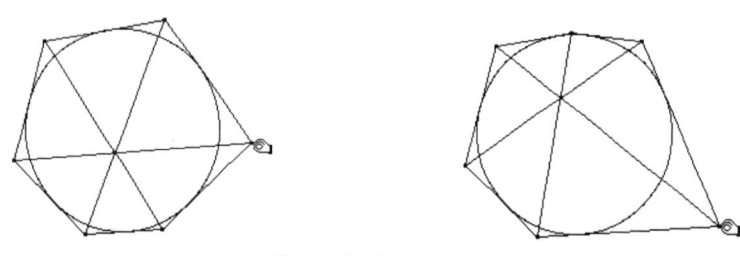

Figs. 10.82–10.83

Example 5 *Cyclic hexagon*
In a cyclic hexagon, the points of intersection of each pair of opposite sides lie on a straight line (Figs. 10.84–10.85) – Pascal's Theorem for the circle as a particular conic section (dual to Brianchon's Theorem).

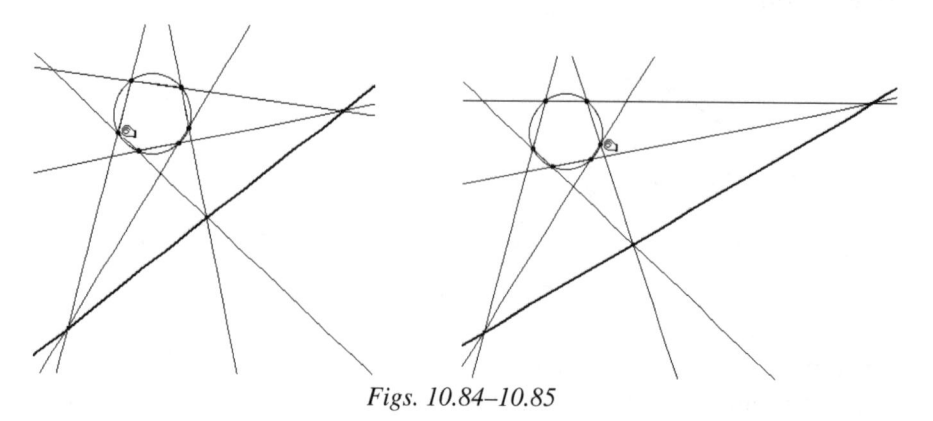

Figs. 10.84–10.85

Exploring drag-mode geometry

Example 6 *A further hexagon theorem*

We start by constructing an aesthetically pleasing configuration using regular hexagons. We begin with a regular hexagon as base (bold in Fig. 10.86). Another regular hexagon (called the midpoint hexagon) has as its vertices the midpoints of the sides of the base hexagon. Further regular hexagons are drawn on the sides of the midpoint hexagon. The midpoints of each pair of oppositely positioned outer hexagons are linked; these three equal length connecting line segments intersect each other at 60° at one point (Fig. 10.86). Now one vertex of the base hexagon is released from its construction connection (using **Redefine an object** or **Delete relation**) and dragged away (Fig. 10.87, generalisation). We watch the connecting line segments; they remain of equal length (Figs. 10.88–10.89), but no longer intersect at one point, although the angle of the intersections does remain 60°.

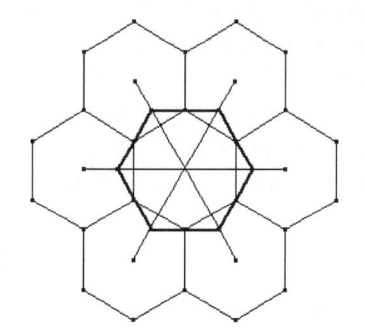

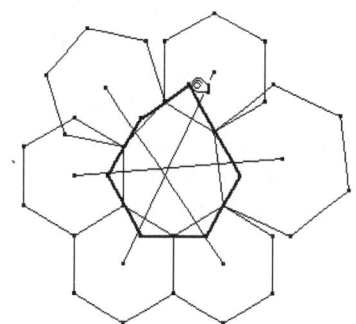

Figs. 10.86–10.87

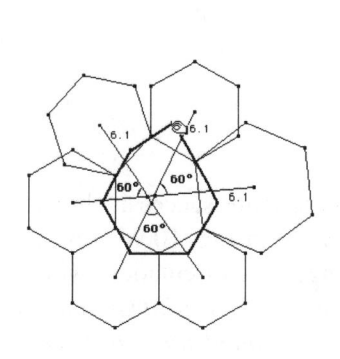

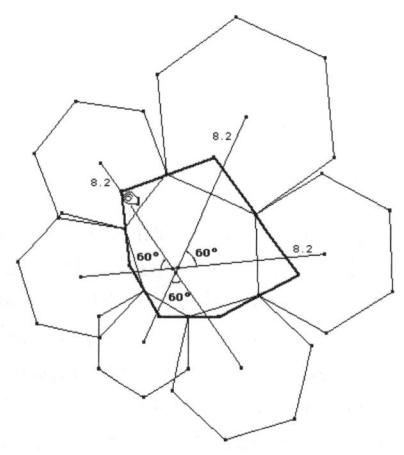

Figs. 10.88–10.89

185

Discovering Geometry with a Computer

This also applies to the non-convex case (not illustrated). These properties do not carry over to octagons (Figs. 10.90–10.91).

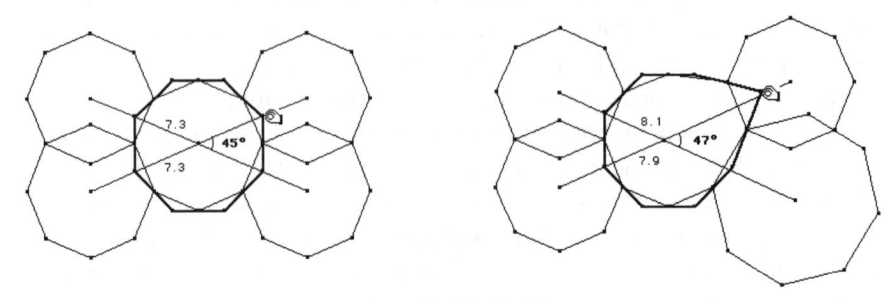

Figs. 10.90–10.91

We conclude this section with examples investigating combinations of transformations. In Fig. 10.92 point reflections of arbitrary points A and B (defining a line segment) are successively executed in two points P_1, P_2, producing ultimately A" and B", which are joined. It is found that A"B" and AB remain parallel as point B is moved (Fig. 10.93), which indicates a translation, and this is confirmed when any other point is moved (Fig. 10.94: point P_2).

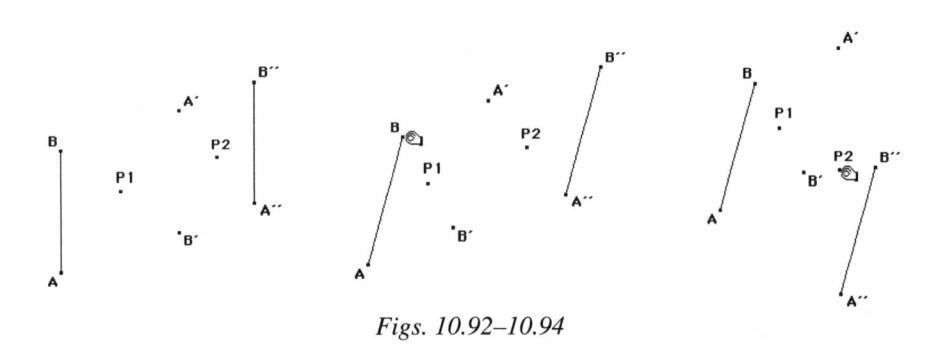

Figs. 10.92–10.94

In Fig 10.95 point reflections of an arbitrary point P are successively carried out in four points P_1, P_2, P_3, P_4, Moving P (as in Fig. 10.96) reveals that the connecting line PP' retains the same length and orientation, which indicates a translation. An alternative interpretation is: *four arbitrary points can generally not be the midpoints of the sides of a quadrilateral.* Where must these four points lie so that P=P', i.e. the product of four point reflections is the identity? (This is the case n=2 of the general statement for 2n points.)

Exploring drag-mode geometry

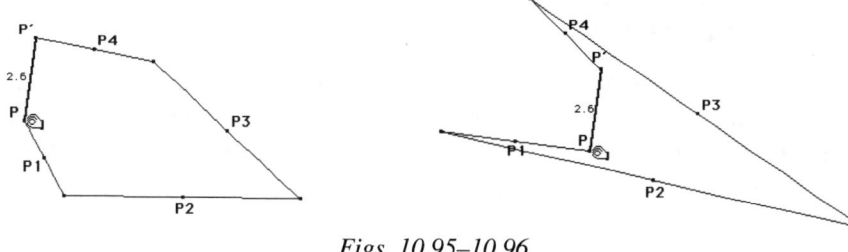

Figs. 10.95–10.96

With five points P_1, P_2, P_3, P_4, P_5 (generally $2n + 1$ points), the midpoint of PP' is fixed as P varies, i.e. the product of the $5(2n + 1)$ point reflections must also be a point reflection (Figs. 10.97–10.98). $5(2n + 1)$ points can always be midpoints of the sides of a pentagon (in general a $(2n+1)$-gon)); the initial vertex P ($= P'$) of the polygon is to be chosen as the centre of the product of transformations (Fig. 10.99).

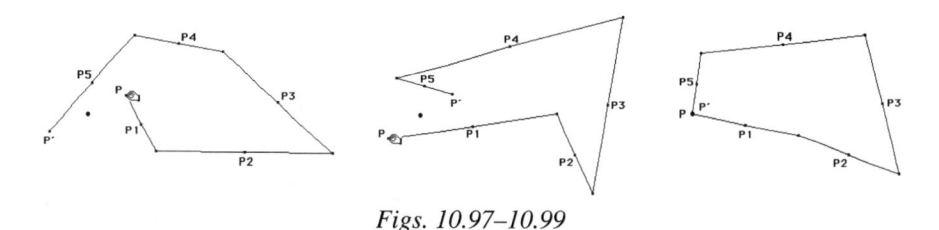

Figs. 10.97–10.99

In the consecutive execution of a point reflection (in Z) followed by a straight line reflection (in g), we discover that the midpoint of PP' lies on a fixed line through Z perpendicular to g (Figs. 10.100–10.102; Fig. 10.102 with the fixed line as an interactively produced locus). That suggests a glide reflection along this fixed line, which is not a fixed-point line (i.e. the overall line maps onto itself but each point does not map onto itself).

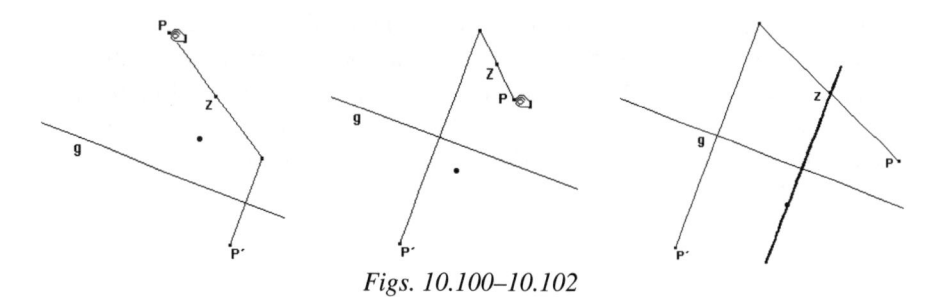

Figs. 10.100–10.102

10.4 ADVANCED APPLICATIONS OF DRAG-MODE TO GENERATE LOCI

A powerful facility within Cabri-géomètre is the **Locus** command. Through the individual dragging of a point, usually along a guide-line, a locus is generated from one (or more) points constructively dependent on the dragged point. Here we give some rather advanced examples, mainly from conic section geometry.

Example 7 *Three ways of generating ellipses*
(a) Method 1
The endpoints A, B of a line segment of a pre-given copied length (shown below the main figure in Fig. 10.103) slide along two orthogonal straight lines (Fig. 10.103; here A can be dragged and B is dependent). Then a point P is placed on this line segment, dividing it into arbitrary but fixed proportions. P is reflected in one of the orthogonal straight lines giving image point P'. As A is moved along its constraining line, P and P' together describe an ellipse (Fig. 10.104).

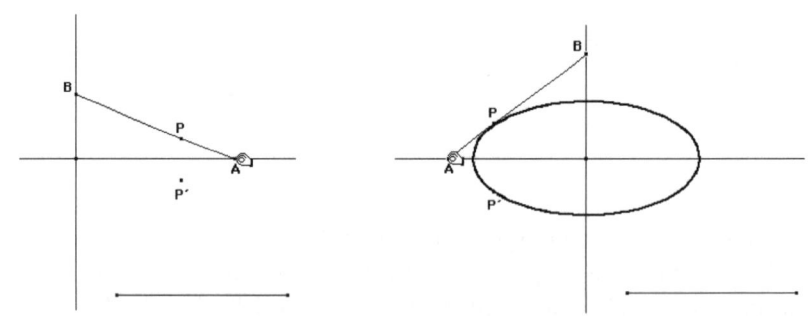

Figs. 10.103–10.104

(b) Method 2
We are given a line segment which is subdivided into two sub-segments by a moveable point on it (shown to the right of the main figure in Fig. 10.105). These two sub-segments are taken as the radii of circles centred on focal points F_1 and F_2. The circles intersect at P and P'. As the free point is moved along its line segment, P and P' generate an ellipse (Fig. 10.106). (Note: the sum of the distances of PF_1 and PF_2 is constant.)

Exploring drag-mode geometry

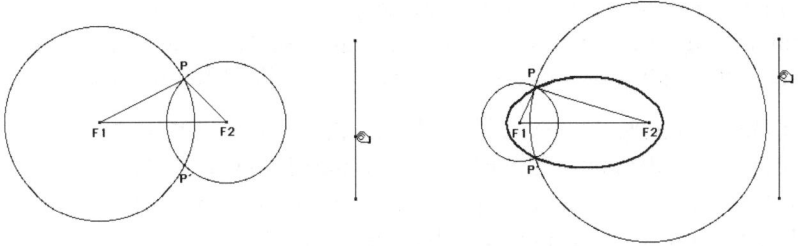

Figs. 10.105–10.106

(c) Method 3

In setting up Example 5 of Section 10.3.2, we used the converse of Pascal's Theorem. The converse states: "If the opposite sides of a hexagon intersect each other on a straight line, its six vertices lie on a conic section". This can be used to projectively construct an ellipse as a locus.

Thus we take five points 1, 2, 3, 4, 5 and put a moveable straight line through point 5 (shown by the grasping hand) whose point of intersection with the straight line through point 1 and point S is denoted as point 6 (Fig. 10.107). By rotating the line passing through points 5 and 6 around point 5, point 6 describes an ellipse and points 1 to 6 all lie on the locus (Fig. 10.108).

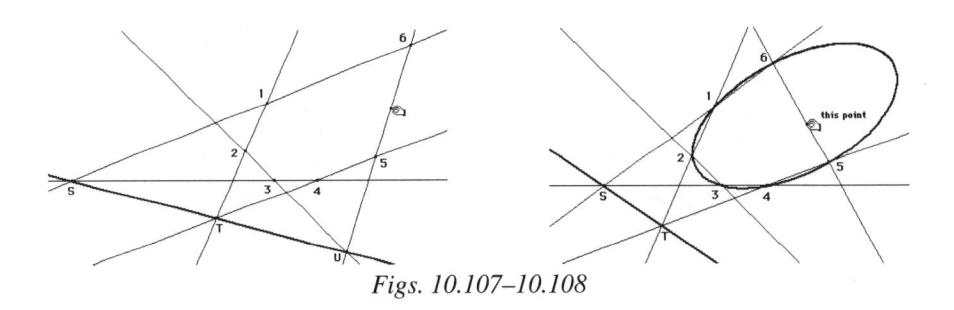

Figs. 10.107–10.108

We could also have used this locus-production to find the converse of Pascal's Theorem.

Example 8a *Pedal curve of circle*
Given a circle centre M and circumference point K, and a separate point D, we draw the radius MK and construct the tangent at K. A line through D

189

parallel to MK is drawn and its intersection with the tangent is marked as point L (Fig. 10.109). We move tangent-point K around the circle and observe the path which L describes. We distinguish different cases:

- If D lies outside the circle, we obtain a Pascal's snail (Fig. 10.110).
- If D lies on the circle, a cardioid is produced (Fig. 10.111).
- If D lies inside the circle, a cardioid is produced (Fig. 10.112).
- If the centre of rotation D and the circle centre K coincide, we obtain the circle itself (not illustrated).

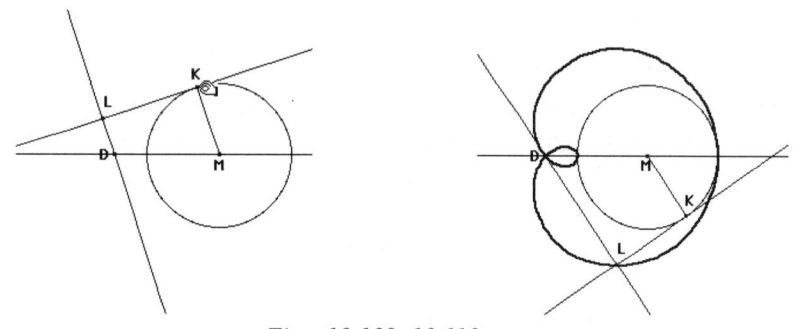

Figs. 10.109–10.110

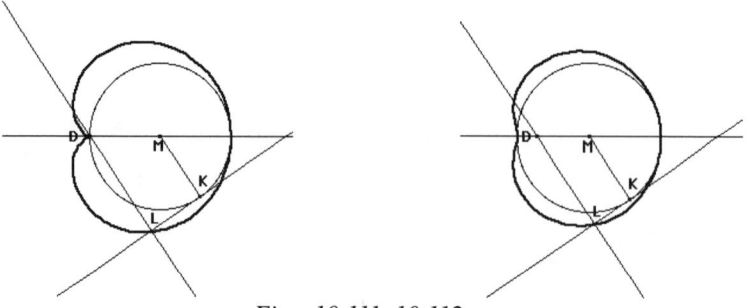

Figs. 10.111–10.112

Example 8b *Pedal curves of ellipse, hyperbola and parabola*
We wish to use a conic section (ellipse, hyperbola or parabola) to produce pedal curve loci. Unfortunately, Cabri-géomètre has no standard function for the production of ellipses, hyperbolas or parabolas, unlike the circle which was employed in Example 8a. We must therefore construct the required conic section for ourselves and simultaneously use it to produce

the requisite curve. I.e. we construct the appropriate conic section as the locus using a guide-curve, and use its tangent-line and the perpendicular to the tangent passing through a centre of rotation to obtain the required intersection point whose locus is sought. The circle is the appropriate guide-curve for constructing the ellipse and hyperbola; for the parabola, we use a straight line.

(a) *Ellipse*

If the position of the centre of rotation D lies outside the ellipse on its main axis, we obtain a Pascal's snail (Figs. 10.113–10.114).

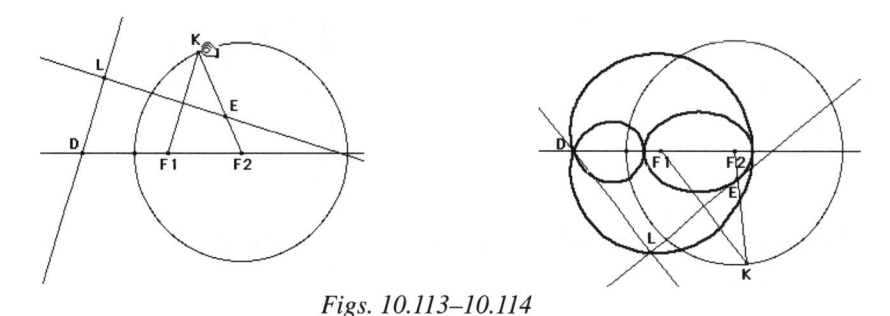

Figs. 10.113–10.114

(b) *Hyperbola*

Using an hyperbola, a lemniscate of Bernouilli is produced if the centre of rotation coincides with the midpoint of F_1F_2 (Figs. 10.115–10.116).

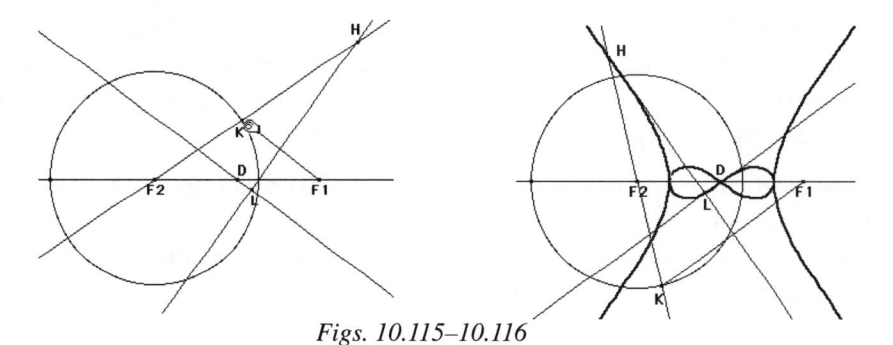

Figs. 10.115–10.116

(c) *Parabola*

The strophoid is the pedal curve of the parabola if the centre of rotation D and focus F lie symmetric to the directrix of the parabola (Figs. 10.117–10.118).

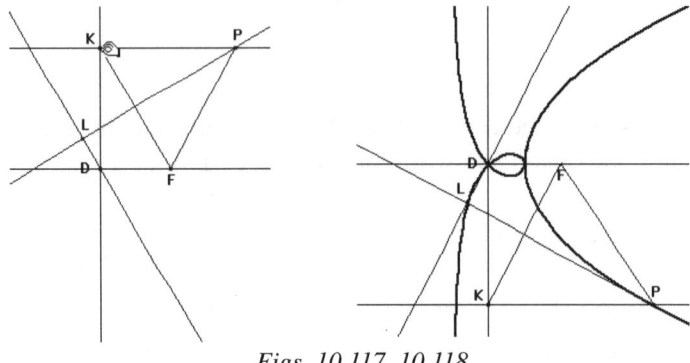

Figs. 10.117–10.118

With the indirect generation of loci we possess a very flexible construction and exploration tool.

Teaching note: In mathematics education until now, algebraic curves (with the exception of conic sections) have not found their rightful place in the syllabus, despite their variety, beauty and applicability and potential to go from synthetic to analytical representation, which is the reverse direction to the conventional teaching mode. A reason for this is the difficulty of generating such curves (special mechanical instruments have found almost no place in school mathematics). With the interactive generation of loci, a new methodological approach arises, for students aged 16+ say, working in groups. Some suggestions are:

(a) Interactive generation of algebraic curves (with an printout for the safeguard of the results!).
(b) Analytical description of the algebraic curves produced (using an appropriate co-ordinate system).

The next example shows that proposed constructions based on an analytical description alone can contain mistaken ideas of the actual generation of such curves.

Example 9 *Scarab beetle curve*

We begin with an translated extract from a German student textbook:

> "With reference to Fig. 10.119 we choose a fixed point A on the y=x line and moveable points P and Q on the x and y axes. A has a distance d from the origin 0. We drop a perpendicular from A onto PQ whose foot is K. As PQ slides on the axes, the point K describes the scarab beetle curve."

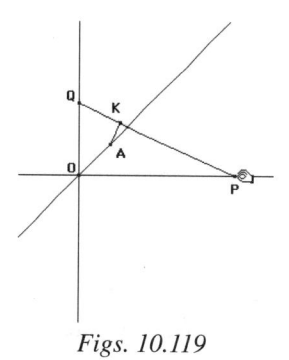

Figs. 10.119

If we follow the above instructions, we obtain only the first branch of the curve. Fig. 10.120 shows part of the first branch and Fig. 10.121 shows the completed first branch.

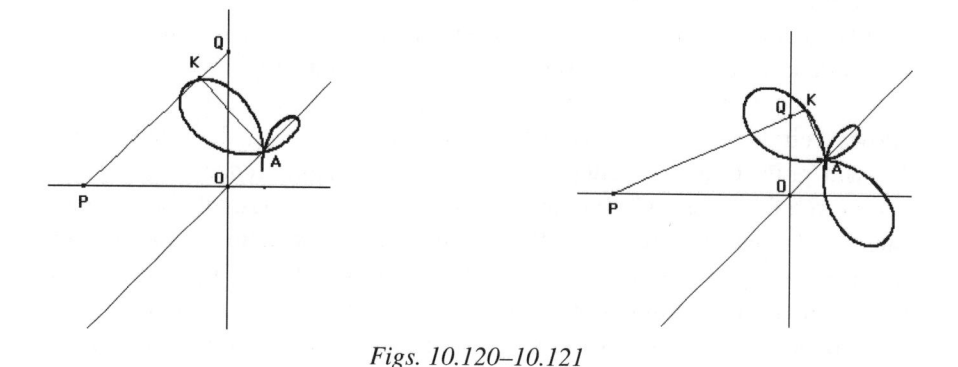

Figs. 10.120–10.121

We need to amend the last sentence and add further instructions as follows:

> "As PQ slides on the axes, the point K describes the first branch of the scarab beetle curve (Fig. 10.121). We reflect PQ in O and drop the perpendicular from A onto P'Q' whose foot is K'. As P'Q' slides on the

axes, the point K' describes the second branch (Fig. 10.122). Thus K and K' generate the whole curve (Fig. 10.123)."

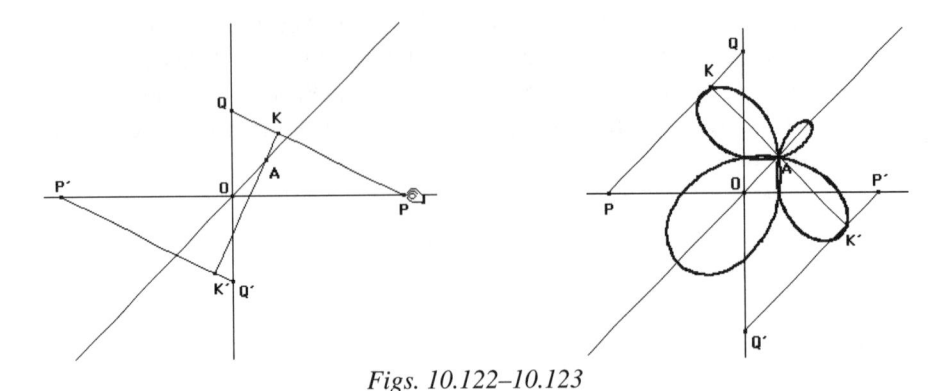

Figs. 10.122–10.123

Apart from introducing an interesting curve, this example shows that having an interactive system such as Cabri-géomètre can bring insight into complex curves which are too difficult to draw by traditional means.

We conclude our rather subjective selection of examples of the interactive production of loci with a classic example from the study of gears – a mechanical device to convert circular motion into linear motion.

Example 10 *Peaucellier's Cell*
The Peaucellier Cell is based on a property of circle inversion: the inverse of a circle which passes through the centre of the inversion circle is a straight line. (This was met earlier in Chapter 7, Section 7.2.3.) Fig. 10.124 shows a circle of inversion, centre O, inside which is a smaller circle centre M with point O and a general point P on its circumference. Fig. 10.125 shows in bold the straight line which is the inverse of the small circle, generated as P moves round it. Fig 10.126 shows the construction of the cell (or invertor), which depends on the length of a moveable rod and radius of a circle (equal to the crank MP) which passes through O. (These two lengths are shown as line segments to the left of the figure; their adjustment can be used to alter the lengths in the figure itself.)

Exploring drag-mode geometry

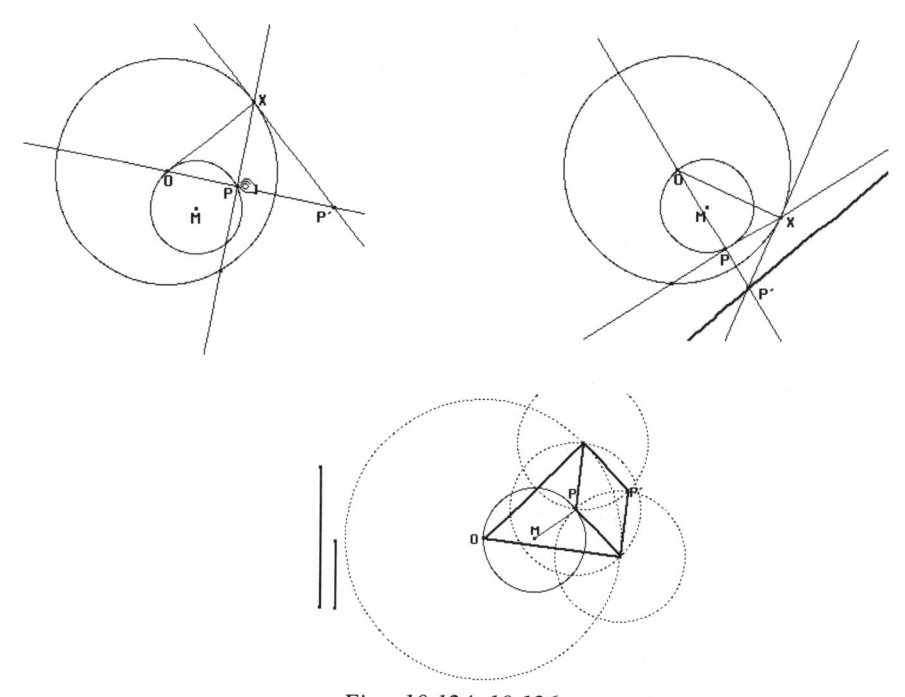

Figs. 10.124–10.126

We can simulate the operation of the invertor by rotating the crank MP about M by dragging P and observing P' (Fig. 10.127 – auxiliary lines omitted). P' generates a line segment (Fig. 10.128 – locus showing *linear* motion as P performs *circular* motion).

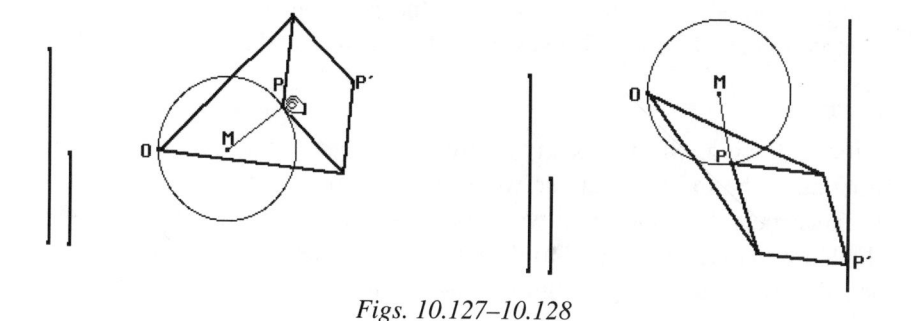

Figs. 10.127–10.128

Further notes.
In general under inversion a circle is mapped onto a circle but if the circle to be inverted passes through the centre of the inversion circle then the

195

image is a circle with infinite radius i.e. it is a straight line. That is the principle of this invertor and explains why O must lie on the small circle. How do we construct the image of the circle with P on it? We know it is a straight line. The following instructions will explain this, to be read in conjunction with Fig. 10.124.

Draw a circle centre M radius |MO| and place on it an arbitrary point P. Construct the circle of inversion centre O with radius > |MO| so that it contains completely the first circle, which is to be inverted. Construct OP which when extended passes through P', the point we wish to find. Erect at P a perpendicular to OP – this intersects the circle of inversion at X. Draw a radius from O to X. Erect at X the perpendicular to the radius OX. P' lies on this perpendicular through X and on OP extended.

P' describes a straight line as P moves round the circle. As P tends to point O on the circle, P' tends to infinity. Figs. 10.127–10.128 show the completed apparatus in action.

In this same way, numerous mechanical devices can be geometrically modelled and their functioning simulated.

10.5 USING DRAG-MODE TO SOLVE CONSTRUCTION TASKS

With drag-mode it is possible to test the universal validity of a solution by checking that the solution is invariant when the given basic objects of the configuration are repositioned. Varying the position of basic objects of a construction can be used to investigate its structure and internal dependencies.

In the following, we do not give a full treatment of the algorithmic construction of the solution. Rather, we restrict ourselves to the development of an experimental solution and we limit ourselves so-called insertion tasks, which are particularly suited to drag-mode. For this we utilise macros and loci. The heuristic strategy employed here is to omit certain conditions.

Task 1: We are given a quadrilateral ABCD and a line segment P_1 whose endpoints should lie on neighbouring sides of the quadrilateral (Fig. 10.129). We seek a parallelogram, inscribed in the quadrilateral, with one side P_1P_2. We produce a third parallelogram vertex P_3 on BC, for example,

Exploring drag-mode geometry

and draw a parallelogram with vertices P_1, P_2, P_3 (Fig. 10.130). We adjust the position of P_3 until P_4 comes to lie on the fourth side of the parallelogram (Fig. 10.131). In order to find a conventional general construction, we generate the geometric locus which P_4 describes when P_3 moves; it is a straight line parallel to BC (note that BC is in fact the guide-line of P_3) (Fig. 10.132).

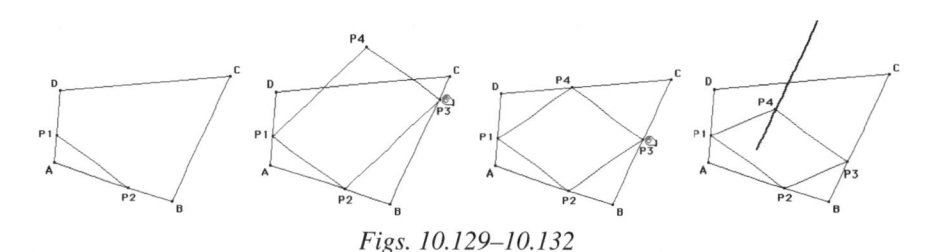

Figs. 10.129–10.132

Task 2: We are given a triangle ABC and a point D on one side of the triangle. We seek an equilateral triangle with vertex D, the other vertices being on AC and BC (Fig. 10.133). We choose any point E on BC and base an equilateral triangle on DE (here we are omitting one condition, namely that the triangle should have its third vertex on the remaining side of triangle ABC: Fig. 10.134). Then we vary the position of E along BC until the free vertex F comes to lie on AC (this entails a rotation-enlargement of DEF about centre D). In order to achieve a ruler-and-compasses construction, we determine the locus which F describes when E moves along BC (Fig. 10.135). This locus is the line segment BC rotated 60° about D.

The solubility of the task depends on the position of D. If D is in the position shown in Fig. 10.136, there can be no solution.

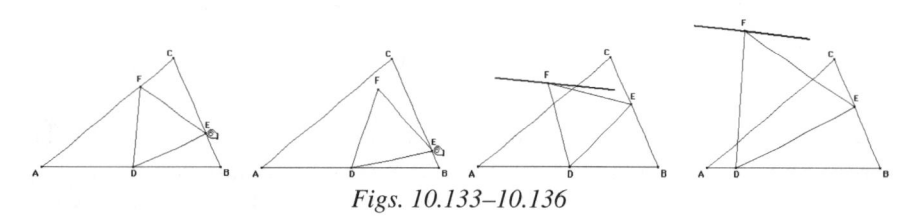

Figs. 10.133–10.136

Task 3: We are given two straight lines g and h and a point P (Fig. 10.137). We seek a square with P as a vertex, which has two other vertices (those neighbouring P) lying on g and h, the opposite vertex having no restrictions. We choose a point A on g and generate a square with side AP (per-

197

haps using the macro SQUARE). The vertex opposite A is labelled B (Fig. 10.138). Now we move A along g until B comes to lie on h and so obtain an experimental solution of the task (Fig. 10.139). As a heuristic strategy to find a ruler-and-compasses construction, we ask for the locus of B as A moves on g (omitting one condition – the need for B to lie on h). The locus produced shows that B moves on a straight line at right-angles to g (Fig. 10.140); the point of intersection of g and the locus line provides the desired position of B on h. P is equidistant from both g and the perpendicular line. Thus the straight line to be constructed (to determine B's position) consists of g rotated through 90° about P.

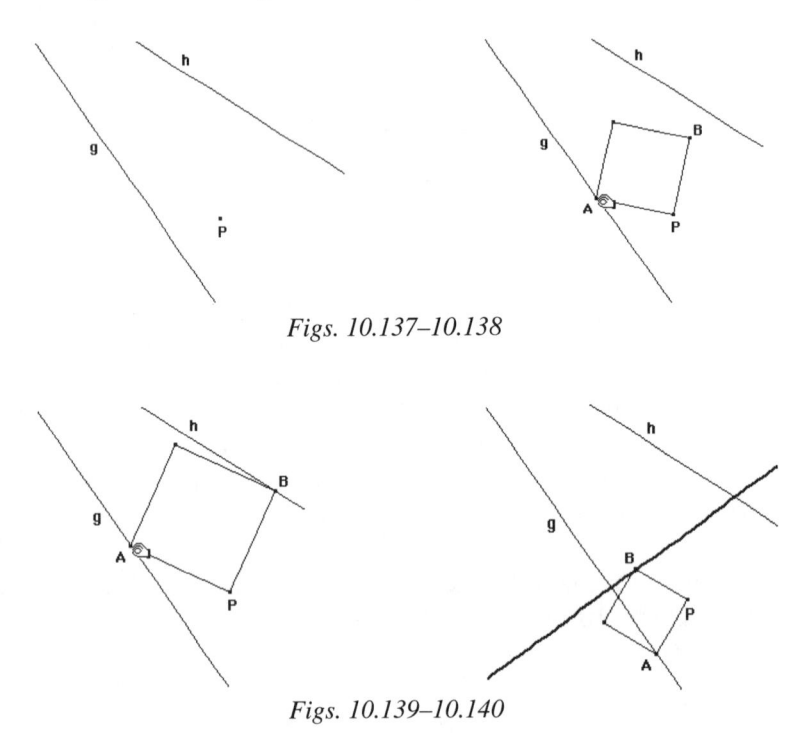

Figs. 10.137–10.138

Figs. 10.139–10.140

Task 4: We are given a straight line g, a point P and a circle k (Fig. 10.141). We seek a line segment with midpoint P, whose endpoints lie on k and g. We choose a point K on k and reflect this in P to produce K'. The desired line segment will be a particular case of KK' (Fig. 10.42). There are two solutions, obtainable by moving K on k (Figs. 10.143–10.144). What is the locus described by K' as K is moved on k? It is the image of circle k under point reflection in P, and its two points of intersection with g are the endpoints of the solution line segments to be constructed (Fig. 10.145). The solubility of the task depends on the position of P in relation to k and g (Fig. 10.146). It is left to the reader to systematically determine what are the solubility conditions.

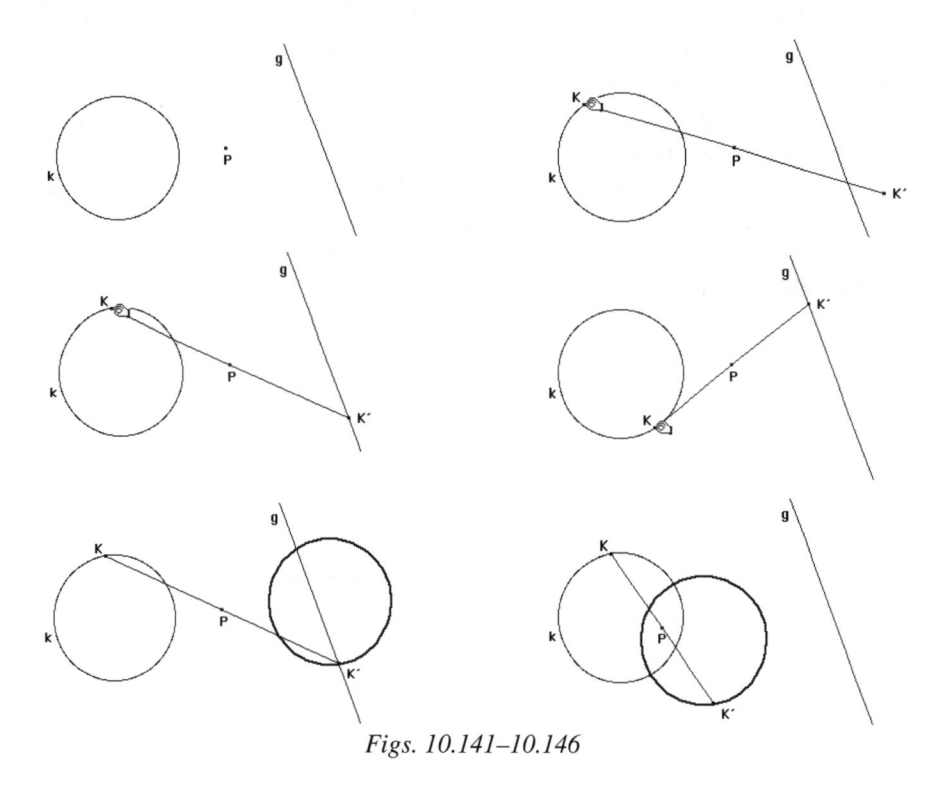

Figs. 10.141–10.146

10.6 DRAG-MODE USED FOR TRANSFORMATIONS IN SCHOOL GEOMETRY

The parameters which determine a diagram can be continuously varied using drag-mode. Such variation can aid initial understanding and study of the properties of mappings. Numerical parameters, such as measurements of angles and lengths, can also be associated with graphic objects on screen. Special and limiting cases of a diagram can be studied. Variation of the positions of determining objects in which reflections are made (points or straight lines) is also possible. By observing figures in motion we can deduce or recognise properties of the underlying transformations. We are not interested here in the figures themselves but rather in what their movements tell us about the transformations.

Discovering Geometry with a Computer

We illustrate the variation of the parameters determining a diagram with an example of axial affinity. In Fig. 10.147 the proportion of affinity is adjustable on a sliding scale placed at the right side of the figure, and the angle of affinity is adjustable through an opening angle of a circle sector. The axis of affinity is a moveable horizontal line (just above the zero position). Adjustment is achieved by dragging a marker point (shown with the grabbing hand icon). In Fig. 10.147 we have moved the marker slightly away from 1 and thereby indirectly have moved triangle A'B'C' out of its position covering triangle ABC (Proportion of affinity = 1 for them to coincide). Further movement leads to degeneration of the triangle to a line (Fig. 10.148, Proportion of affinity = 0) and then further movement causes inversion (Fig. 10.149, Proportion of affinity = −1).

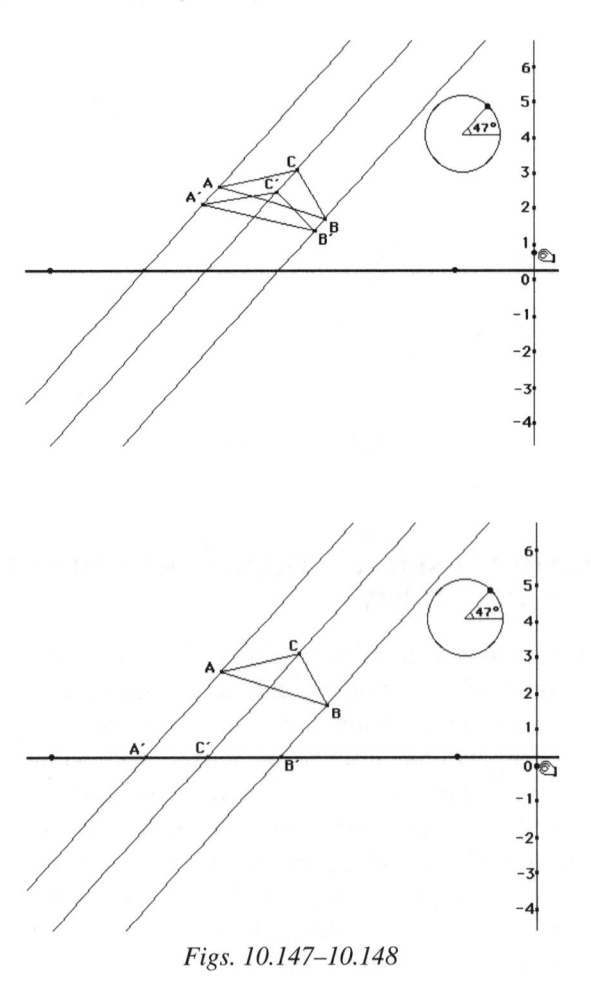

Figs. 10.147–10.148

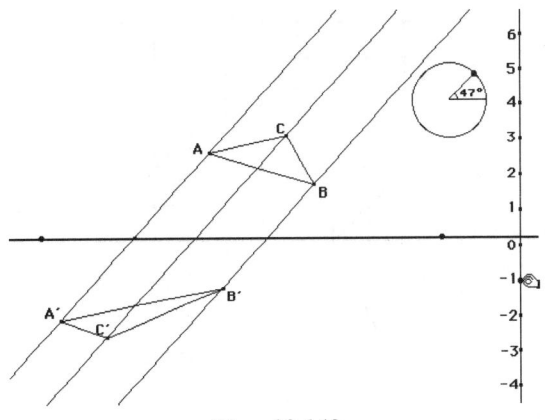

Figs. 10.149

Now we change the angle of affinity to 90° (Fig. 10.150) and obtain a line reflection. Finally, we change the angle *and* vary the position of the axis of affinity so that it is no longer horizontal in relation to the triangle to be reproduced (Fig. 10.151). (The fixed lines could be hidden and would then have to be recognised and drawn in by the user.)

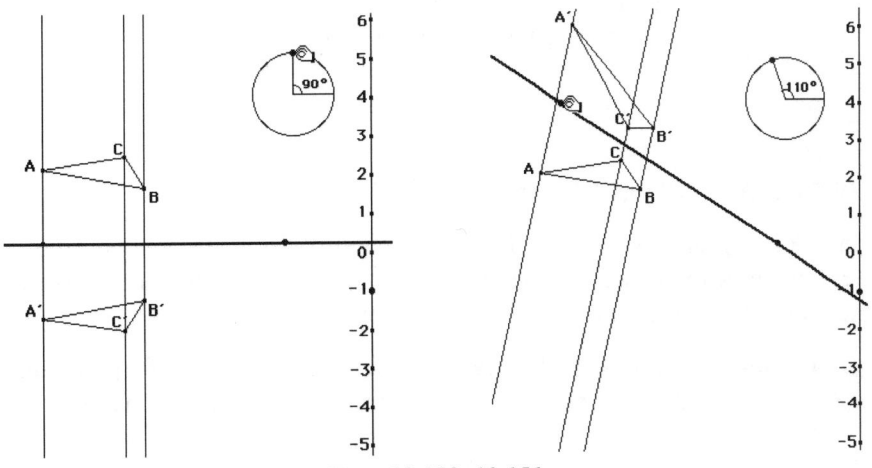

Figs. 10.150–10.151

We close this section with a light-hearted example: A man is drawn in a square grid (Fig. 10.152), which can be stretched or sheared. Fig. 10.153 shows a stretch with scale factor about 0.7. Fig 10.154 shows a shear – the eyes and buttons are invariant under this transformation!

Macros for geometrical transformations are the subject of Chapter 13.

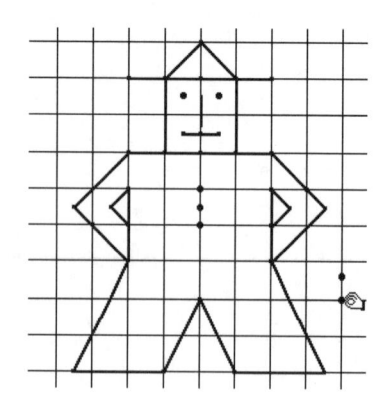

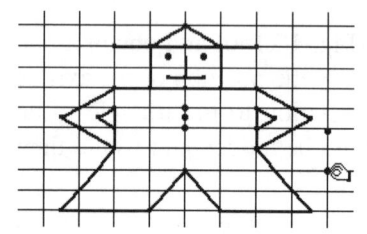

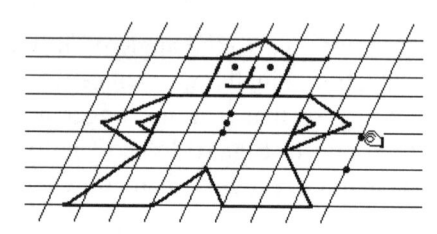

Figs. 10.152–10.154

10.7 REMARKS ON USE IN LESSONS

Interactive graphics systems, like other teaching software, can be used in lessons in different ways – according to the availability and abundance of suitable computers. The following modes of use are suggested:

(a) Demonstration-mode, if only one computer is available (preferably portable and equipped with a screen projection facility to use with an OHP);
(b) Interaction-mode (one computer available for every two students for instance);
(c) A combination of (a) and (b).

The authors have found (c) to be ideal and can report that students' worksheets have proved suitable for efficient working in Interaction-mode. The teacher can decide what constructions to make available to the students on disc for further manipulation and completion.

For various reasons, the use of interactive graphics systems for drawing in school geometry can only supplement the use of the traditional compasses, ruler, set square etc. When a degree of facility in handling **conventional** instruments has been achieved then the time is ripe for **interactive computer** work. There are many demands on a school's hardware today, although in ten year's time personal computer notebooks may well be the norm, which will have a profound effect on mathematics education, geometry included.

CHAPTER 11

Designing geometrical microworlds

The purpose moulds the tool.

11.1 INTRODUCTION

The general goal of designing 2-D geometrical microworlds for school mathematics is to increase access to geometric facts and understanding through a reduction of the demands made on the student to construct figures. An essential feature is a flexible interface between system and user – through menus and the direct manipulation of the graphical configurations. Ideally, the microworld can be *adapted* to vary the geometric objects to be studied and so be tailored to the learning needs of the particular student. These requirements are met to a great extent by the prototype system Cabri-géomètre, which continues to be developed further. (We do not address here the possibilities which intelligent tutorial systems might offer in the future.)

What is needed is a system which can be customised to the student's needs. The following are desirable customisation features of microworlds:

(1) *Customisation of menus*
Purpose: Adaptation of the system to a particular age group, learning style, ability level, current knowledge, familiarity with the tool, constructional complexity of problems, etc., (along the continuum novice – expert).
(2) *Advanced preparation of graphical configurations* for students to manipulate or extend.
Purpose: Experimental solution of individual geometric problems.

Examples of different types of microworlds for secondary level students are presented in this chapter. Experience has shown that the weaker student often fails in the construction of a geometric configuration which is essential for the solution of the underlying geometric problem – for example a measuring problem or a problem of theorem finding. By using Cabri-géomètre, we can reduce the demands on the student to a minimum by supplying prepared configurations and simplifying or enhancing menus appropriately.

The design of microworlds by means of menu modification and figure preparation is a conceptual challenge for the teacher of geometry. Alternatives to the teacher designing the system alone are design by the student alone or in cooperation with the teacher or buying a ready-made system. The complete microworld, however, always include the teacher, who remains indispensable as organiser, tutor and expert in geometry.

11.2 DESIGN OF MICROWORLDS THROUGH MENU REDUCTION

11.2.1 Simplification

To make it easier for the novice first meeting Cabri-géomètre, the system's range of options in the **File**, **Edit**, **Creation**, **Construction** and **Miscellaneous** menus can be reduced using the **Edit menus** option.

The type of reduction can be determined by the students' age and previous knowledge of geometry and didactical aims.

Example 1
This menu design (Figs. 11.1–11.5) is suitable as a first system for 11 year olds.

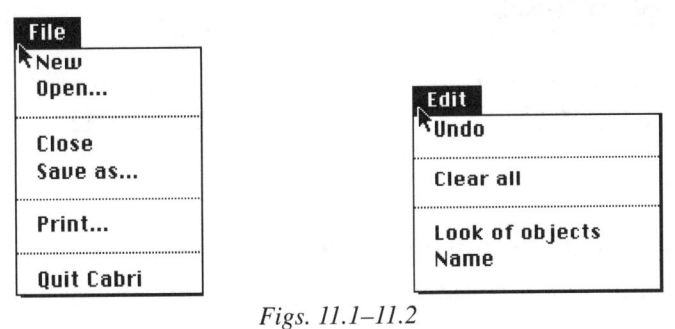

Figs. 11.1–11.2

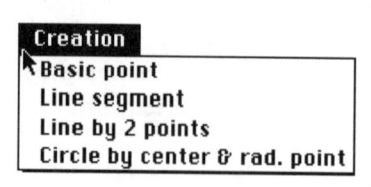

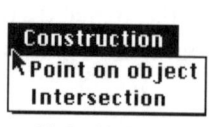

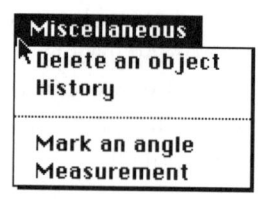

Creation
* Basic point
* Line segment
* Line by 2 points
* Circle by center & rad. point

Construction
* Point on object
* Intersection

Miscellaneous
* Delete an object
* History
* Mark an angle
* Measurement

Figs. 11.3–11.5

Example 2

This menu design (Figs 11.6–11.10) is suitable as a first operational system for 11 year olds who have received an elementary introduction to Cabri-géomètre. It allows the student to extend the construction menu by designing and defining macros to supply basic constructions (which could be compared with the corresponding system macros).

File
* New
* Open...
* Close
* Save as...
* Revert
* Print...
* Quit Cabri

Edit
* Undo
* Clear all
* Look of objects
* Enlarge the figure
* Name

Creation
* Basic point
* Line segment
* Line by 2 points
* Circle by center & rad. point

Figs. 11.6–11.8

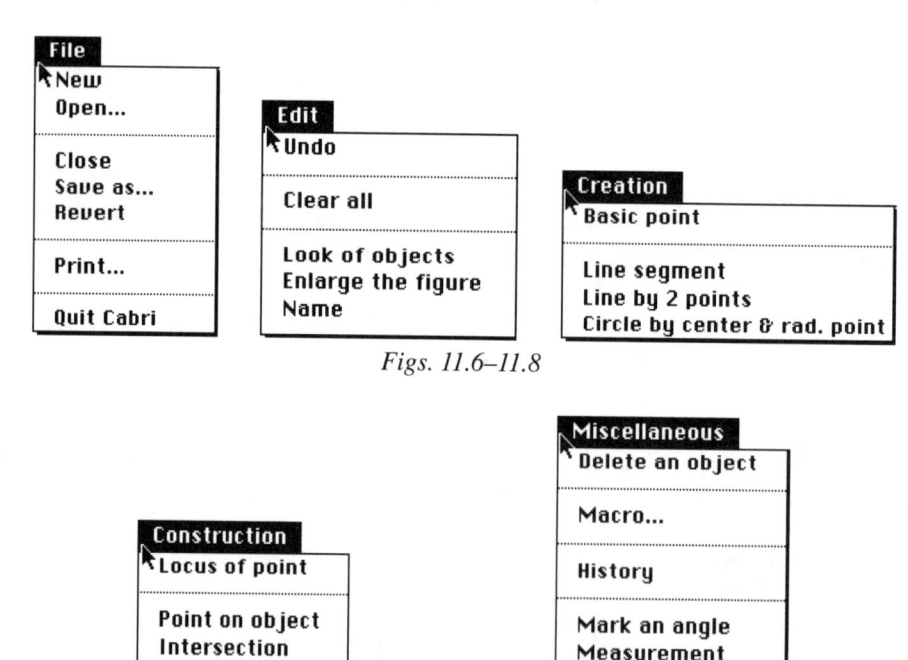

Construction
* Locus of point
* Point on object
* Intersection

Miscellaneous
* Delete an object
* Macro...
* History
* Mark an angle
* Measurement

Figs. 11.9–11.10

11.2.2 Restriction of constructional tools permitted

The goal of such microworlds is to solve constructional problems through a *restriction of the drawing tools*. (The theory of geometric constructions addresses the following central question: Which constructional problems can be solved with which constructional instruments? For example, G. Mohr proved in 1672 that all problems which can be solved with ruler and compasses can be solved with compasses alone; thus all linear objects and properties in Cabri-géomètre menus might be eliminated from an initial construction system. Likewise, a large number of geometric problems can be solved with the ruler alone.)

In Cabri-géomètre's initial default setting, the system is designed for constructions with compasses and ruler; in addition, constructional results can be varied continuously using drag-mode. In the following, we indicate three possibilities for restricting constructional tools.

Example 3 *Ruler-only construction*
The **Creation** menu and the **Construction** menu are restricted as shown in Figs. 11.11–11.12.

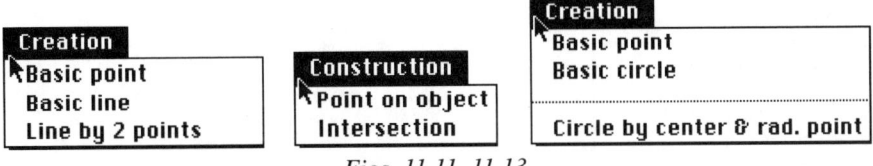

Creation		Creation
Basic point	Construction	Basic point
Basic line	Point on object	Basic circle
Line by 2 points	Intersection	Circle by center & rad. point

Figs. 11.11–11.13

Example 4 *Compasses-only construction*
The **Creation** menu is as shown in Fig. 11.13 and the **Construction** menu as shown in Fig. 11.12.

The command **Symmetrical point** in the **Construction** menu allows the simulation of certain set-square constructions.

11.3 SUPPLYING SPECIALISED CONSTRUCTION MENUS

The student loads the graphics system tailored to the topic area with customised **Creation** and **Construction** menus (or, with the guidance of the teacher, the students might create the microworld themselves).

On-line help supplied with the definitions of the macros can assist correct use of the macros.

Discovering Geometry with a Computer

Example 5 *Triangle constructions*

Fig. 11.14 shows a **Construction** menu for those triangle constructions usually dealt with at school level. The macros CIRCLE-C-r, COPY-LENGTH, COPY-ANGLE (described in Chapter 3) refer to the corresponding ruler-and-compasses constructions, since with Cabri-géomètre the input of measurements via the keyboard is impossible. Line segments or angles which are parameters for the configuration are drawn to one side and copied to form the configuration itself. These lengths and angles can be adjusted using the mouse. As an example, the help explanation of the macro COPY-LENGTH is given in Fig. 11.15.

Construction
Locus of point
Point on object Intersection
Midpoint Perpendicular bisector Parallel Line Perpendicular line Center of a circle
Symmetrical point Bisector
CIRCLE-C-r *COPY-ANGLE* *COPY-LENGTH* *PERP-BISECTOR* *ANGLE-BISECTORS* *CIRCUMSCRIBED-CIRCLE* *INSCRIBED-CIRCLE* *ALTITUDES* *MEDIANS* *MIDSEGMENTS*

COPY-LENGTH:

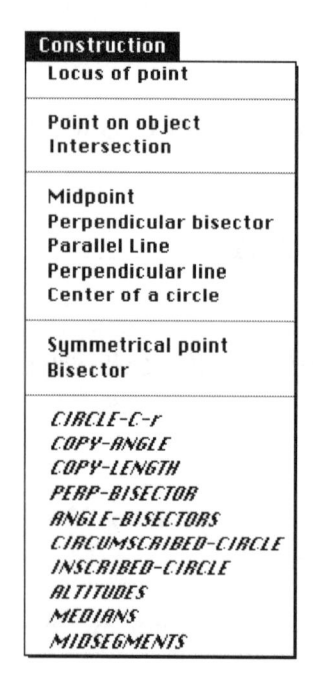

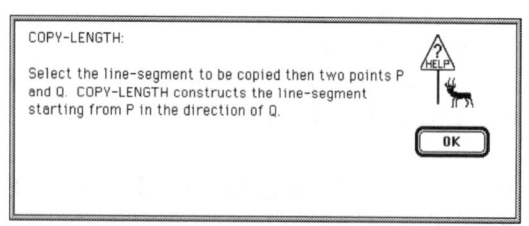

Select the line-segment to be copied then two points P and Q. COPY-LENGTH constructs the line-segment starting from P in the direction of Q.

Figs. 11.14–11.15

Example 6 *Congruence transformations*

Fig. 11.16 shows a **Construction** menu for congruence transformations, in particular enabling their combination. The student's task is to complete a combination table (Fig. 11.17). The parameters of the transformations can be varied and the effect on a (variable) triangle observed. Macros are explained by on-line help (Fig. 11.18, explanation for the 'glide reflection').

Designing geometrical microworlds

Construction		O	T	R	L	G
Locus of point		T				
Point on object Intersection						
Midpoint Perpendicular bisector Parallel Line Perpendicular line Center of a circle		R				
Symmetrical point Bisector		L				
TRANSLATION *ROTATION* *LINE-REFLECTION* *GLIDE-REFLECTION*		G				

GLIDE-REFLECTION:

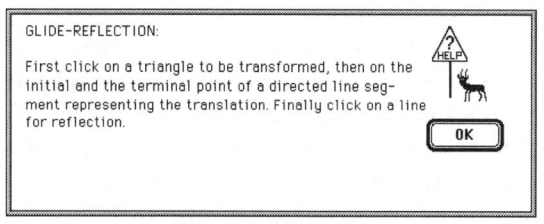

First click on a triangle to be transformed, then on the initial and the terminal point of a directed line segment representing the translation. Finally click on a line for reflection.

HELP

OK

Figs. 11.16–11.18

Example 7 *Circle tangents*

Fig. 11.19 shows a **Construction** menu to tackle the following problem (using circle inversion): Which circles through P touch circles k1, k2 (Fig. 11.20)?

This problem can lead to the more general **contact** problem of Apollonius (in which the circles touching three pre-given circles are to be constructed). Because Cabri-géomètre has no way to choose between two courses of action, the distinction between circle reflection from inside the inversion circle to outside it (I – O) and from outside the inversion circle to inside it (O – I) must be made explicit.

```
Construction
  Locus of point

  Point on object
  Intersection

  Midpoint
  Perpendicular bisector
  Parallel Line
  Perpendicular line
  Center of a circle

  Symmetrical point
  Bisector

  TANGENTS-TO-2-CIRCLES
  3-POINT-CIRCLE
  INVERSION 1-0 circle
  INVERSION 1-0 point
  INVERSION 0-1 point
  INVERSION 0-1 line
```

Figs. 11.19–11.20

The above problem can be reduced to the construction of the common tangents to the inverses of the given circles, whereby the tangents map into the desired contact circles when re-inverted in the inversion circle i with centre P (Fig. 11.21). The centre of the inversion circle must be P but the radius is arbitrary – although it is best to have the two initial circles totally inside the inversion circle.

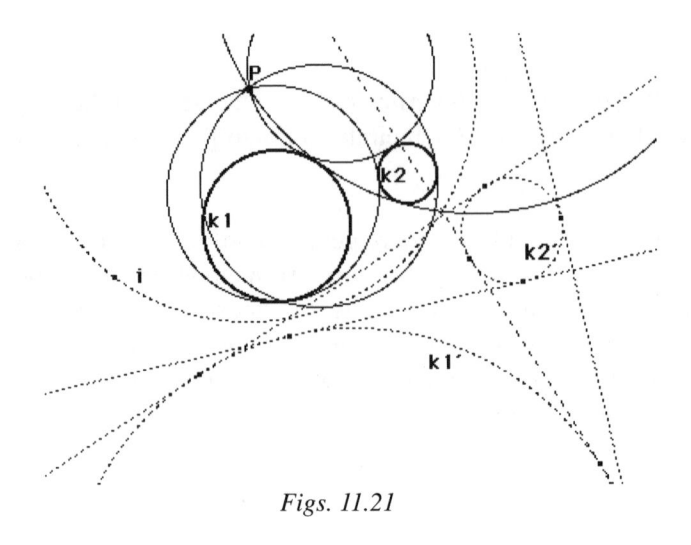

Figs. 11.21

11.4 SUPPLYING PREPARED CONFIGURATIONS

Another use of microworlds is to provide particular configurations, rather than menu commands, as the basis for experimentally solving geometric problems.

The following examples come from school geometry; applied problems have been included and in particular those dealing with the simulation of motion (e.g. mechanical devices). The primary objective is the experimental acquisition of knowledge. The question "How did the person manage to construct this?" should motivate students to try to reconstruct the given configuration.

Example 8 *Frieze pattern*
Problem: Complete the empty fields in Fig. 11.22 by means of reflection of the given triangle and its reflections. Conceal the auxiliary lines and change the shape of the given triangle and its reflections as you wish.
Choose: LINE-REFLECTED-TRIANGLE, **Appearance of object**s

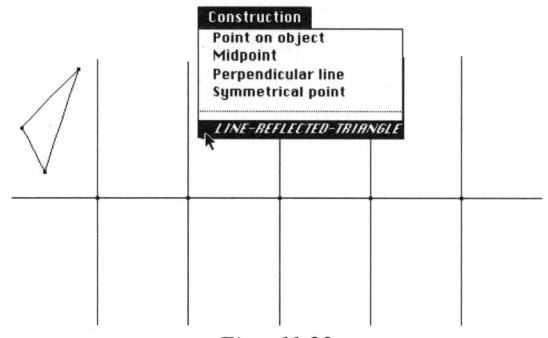

Figs. 11.22

Comment: The generation of representatives of the seven types of frieze pattern is an aesthetically attractive task. Variation of the 'triangle' motif, to produce frieze patterns with symmetrical properties, is encouraged. Figs. 11.23–11.26 show some results. Investigations of the symmetries can be included.

Discovering Geometry with a Computer

Figs. 11.23–11.24

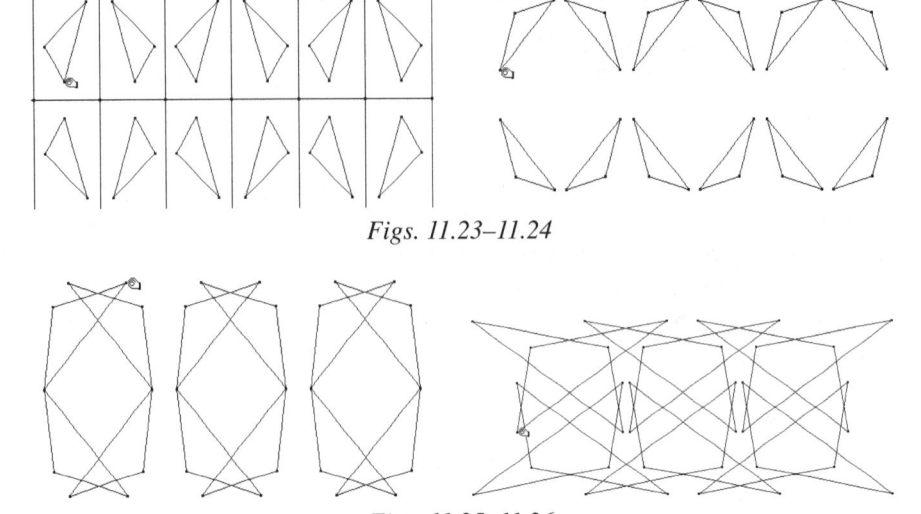

Figs. 11.25–11.26

Example 9 *Production of regular polygons*
Problem: An isosceles triangle is reflected repeatedly in one of its equal sides (Fig. 11.27). What figure do you obtain if you continue and produce a closed figure? Observe the angle at the apex of the isosceles triangle. Modify the figure to a 7-gon, 8-gon, 9-gon, etc.

Comment: The goal is to obtain regular polygons experimentally as closing figures, which requires the angle at the apex of the isosceles triangle to be $360°/n$. Fig. 11.28 shows the closing figure for $n=10$ and Fig. 11.29 shows that for $n=7$.

This method can be used similarly for the production of regular star polygons (for which equidistant circle points are connected).

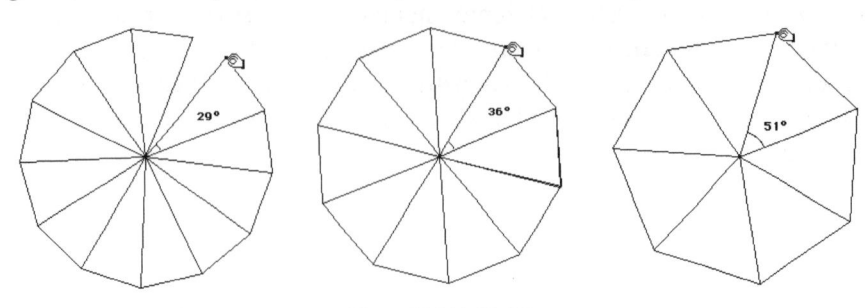

Figs. 11.27–11.29

212

Example 10 *Tessellation*

Problem: A quadrilateral is given (Fig. 11.30), whose shape can be changed by dragging the vertices. Try to tessellate the plane by using one quadrilateral shape. Vary the tessellation. Choose: LAY-DOWN-QUADRILATERAL.

Comment: Plane tessellation problems belong to the class of attractive, easily understood yet sophisticated and partly unsolved mathematical problems. (cf. David Hilbert's Problem 18: to build space using congruent polyhedra; one of Hilbert's world famous 23 research problems presented at the International Mathematics Congress in Paris, 1900). Tessellating the plane with congruent regular 3, 4 or 6 sided polygons, as well as more general quadrilateral tessellations, may be met in school mathematics.

Tessellating quadrilaterals can be achieved with Cabri-géomètre (Fig. 11.31 shows the on-line help).

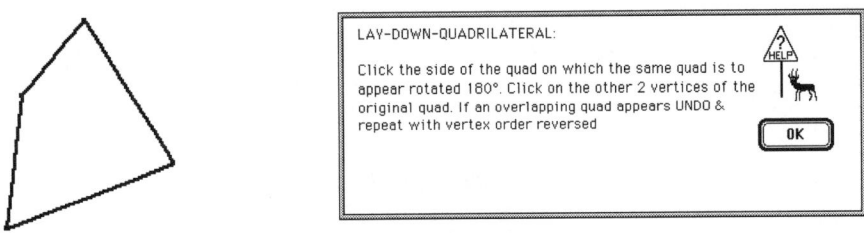

Figs. 11.30–11.31

The triangle is the only basic polygon in Cabri-géomètre, and any macro for constructing a polygon leads to an *n*-point object rather than a single entity (so dragging distorts it, for example). Fig. 11.32 shows part of a tessellation; Figs. 11.33–11.34 show variations obtained by dragging a vertex. Discovering a mapping of the tessellation onto itself, and the theorem of the inner angle sum of the quadrilateral, can be included as student exercises. The interested reader is referred to the excellent book *Tilings and Patterns* by Grünbaum and Shephard published by W H Freeman, 1987.

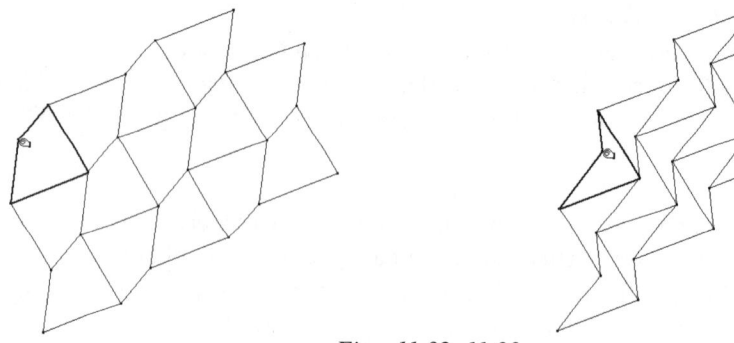

Figs. 11.32–11.33

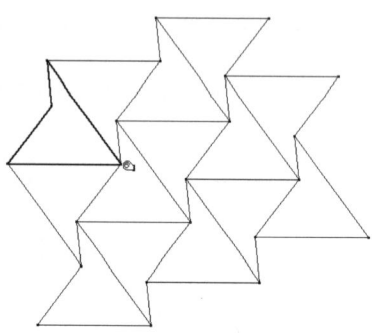

Figs. 11.34

Example 11 *A theorem about point-symmetric hexagons*
Problem: A point-symmetric hexagon is supplied. Equilateral triangles are placed on three sides of the hexagon as in Fig. 11.35. The apexes of these triangles are joined with line segments as in Fig. 11.36.
Modify the point-symmetric hexagon.
What can be ascertained? (If necessary, measure the connecting line segments.)

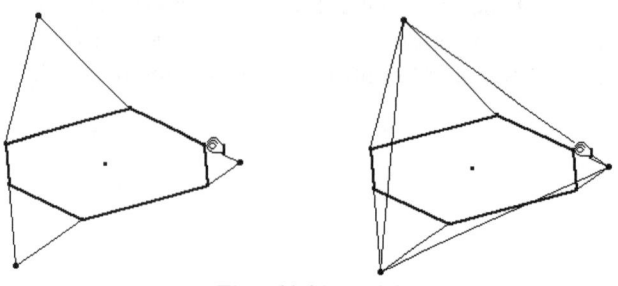

Figs. 11.35–11.36

Comment: This embodies a nice theorem concerning point-symmetric hexagons; the connecting triangle is always equilateral:

- in the case of a non-convex hexagon (Fig. 11.37);
- in the case of a self-intersecting hexagon (Fig. 11.38);
- in the case of a hexagon degenerated to a point-symmetric linkage of three line segments (Fig. 11.39);
- in the case of the equilateral triangle flipping over into the interior of the hexagon (Fig. 11.40).

Requesting verbal formulations of the above invariance statement makes great demands on students.

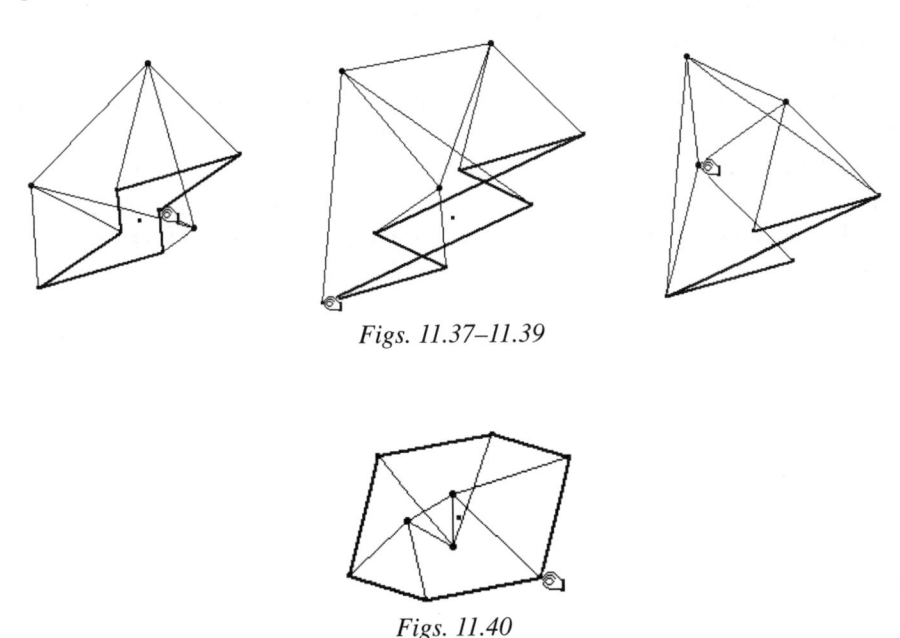

Figs. 11.37–11.39

Figs. 11.40

Example 12 *Minimum distance sum from three points*
Problem: three points A, B, C and a point P are given. The sum of the distances from P to A, B, C can be read from a line segment which is a copy of the three segments (shown below the figure in Fig. 11.41). In addition, the angles APB, BPC, CPA are indicated.

Figs. 11.41

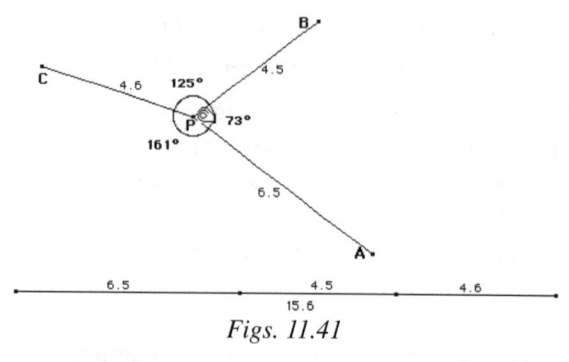

For which position of P is the distance sum least? Check the result for another position of A, B, C; if necessary, draw and measure the angles ABC, BCA, CAB.

Comment: The problem leads to an experimental discovery of the Fermat point for a triangle whose internal angles are each smaller than 120°. Varying the positions of A, B, C may lead to an internal angle of 120° or more, in which case P coincides with the vertex of the largest angle. Fig. 11.42 indicates a reduction of the distance sum compared to Fig. 11.41. The distance sum in Fig. 11.43 cannot be further reduced.

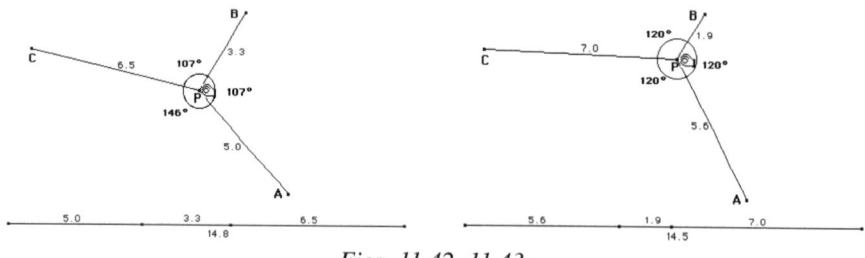

Figs. 11.42–11.43

Example 13 *SSS*

Problem: Three line segments are given, copies of which are used to form a triangle (Fig. 11.44). Vary the length of the given line segments. Can a triangle always be formed? What are the criteria?

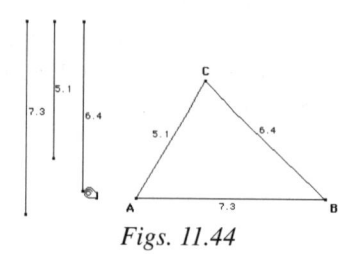

Figs. 11.44

Designing geometrical microworlds

Comment: The dependencies and limits of a construction can easily be investigated experimentally by varying the determining objects. Fig. 11.45 shows the case of a degenerate triangle, and in Fig. 11.46 there is no longer a triangle (Why not?).

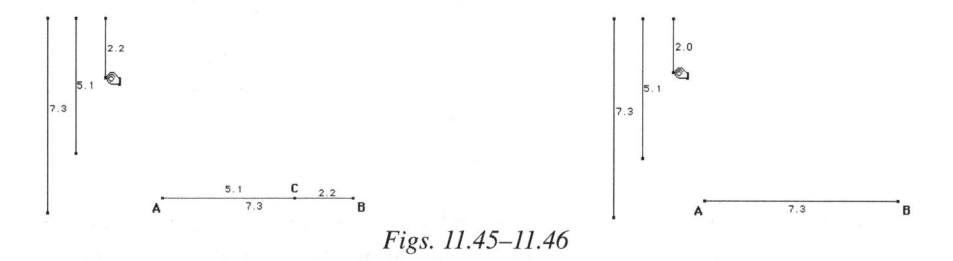

Figs. 11.45–11.46

Example 14 *Isoperimetric triangles*
Problem: The perimeter of the triangle is given as a line segment (in Fig. 11.47 it is 15 units long, shown above the figure). It is divided into three sub-segments which form the three sides of the triangle. Observe the triangle's change in shape as the sub-segments are varied.

The area of the triangle is indicated by a rectangle (in Fig. 11.47 it has height 1 unit and length 5.5 units indicating an area of 5.5 sq units, shown below the figure). Find out for which type of triangle the area is largest.

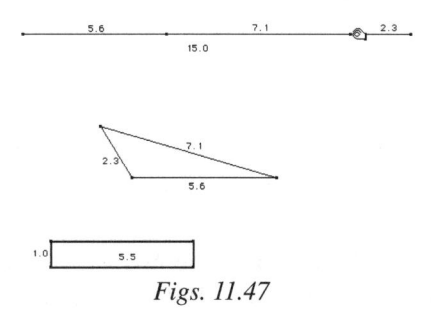

Figs. 11.47

In Fig. 11.48 the triangle's base is held constant and the point of division of the two sub-segments determining the other two sides is varied. The locus of the apex of the triangle is an ellipse: the sum of the two remaining sides is constant; the endpoints of the base are the focal points of the ellipse. A maximum area is obtained for the equilateral triangle (Fig. 11.49).

Comment: Isoperimetric problems are dealt with more fully in Chapter 12.

217

Figs. 11.48–11.49

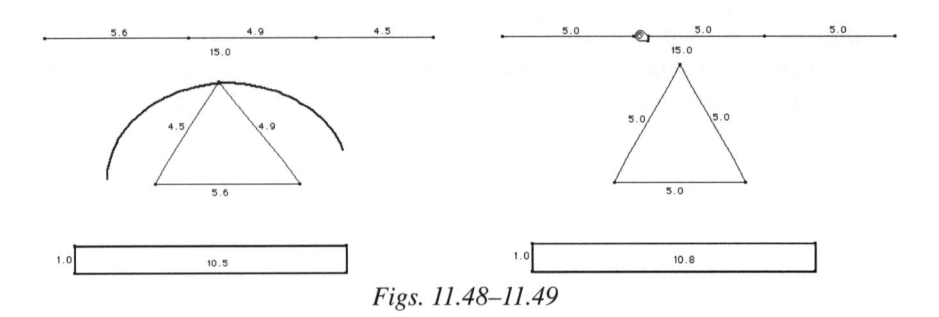

Example 15 *Triangles equal in area*
Problem: The area of a triangle is given by a rectangle with height 1 unit (in Fig. 11.50 the area is 6.9 sq. units, shown above the figure). The shape of the triangle enclosing the given area can be changed by dragging the vertices.

Its perimeter is shown as a line segment (in Fig. 11.50 the perimeter is 13.6 units, shown below the figure). Observe the size of the perimeter when altering the triangle. Find out for which shape the perimeter is smallest.

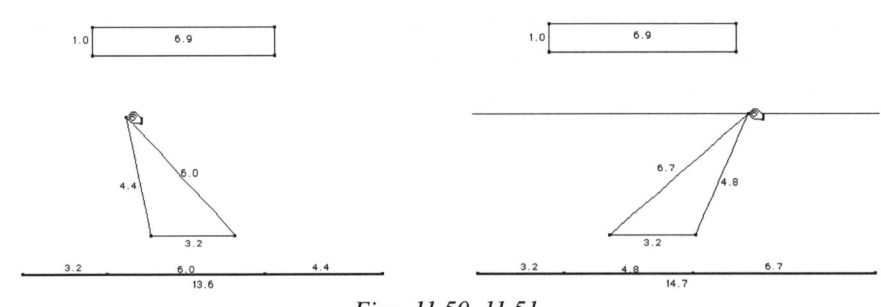

Figs. 11.50–11.51

Comment: This iso-areal problem is the inverse task to the corresponding isoperimetric problem (Example 14). With a fixed base, the apex of the triangle moves on a parallel to the base (Fig. 11.51). The isosceles triangle has the smallest perimeter (Fig. 11.52). Among triangles of equal area, the equilateral triangle has the smallest perimeter (Fig. 11.53).

Designing geometrical microworlds

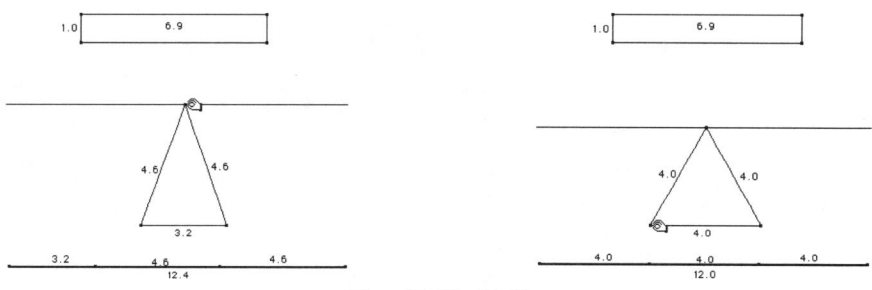

Figs. 11.52–11.53

Example 16 *Enlargement*

Problem: Move the scale point (shown by a gripping hand in Fig. 11.54) and observe triangle A'B'C'. What relationship is there between the sides of A'B'C' and the sides of ABC? Also vary the triangle ABC and investigate.

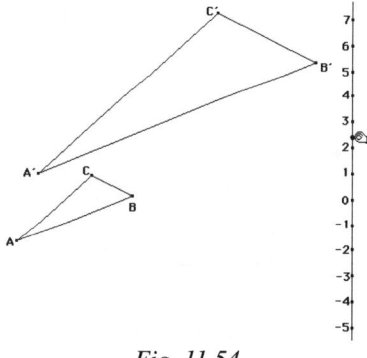

Fig. 11.54

Comment: A triangle ABC is enlarged to form an image A'B'C' by moving the scale factor marker (Fig. 11.54). First, the student should discover the fixed lines and the centre of the enlargement (use locus production as indicated in Fig. 11.55). For special whole-numbered values of the scale factor, a quantitative correlation between image-sides and object-sides is evident (Fig. 11.56), which is experimentally confirmed by varying ABC and the scale factor (Figs. 11.57–11.58). In this way, the figures commonly met in schools can be explored. The student should not have access to the hidden auxiliary lines, but after the problem has been solved the teacher might explain the underlying construction with the help of the **History** command. (See Chapter 13 for more details of such transformations.)

Figs. 11.55–11.56

Figs. 11.57–11.58

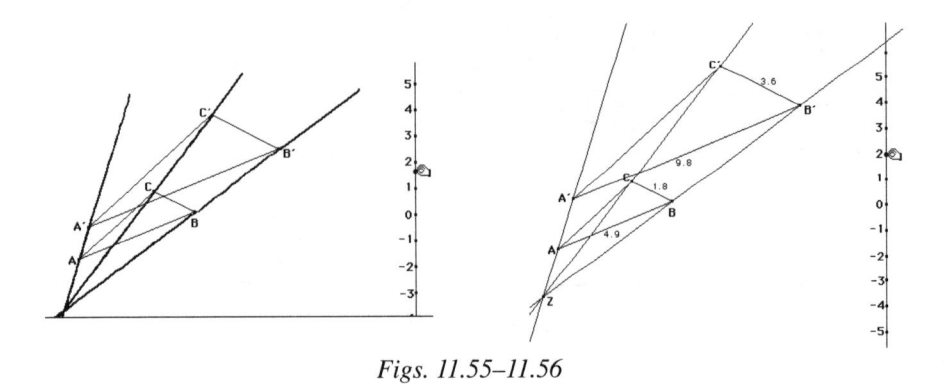

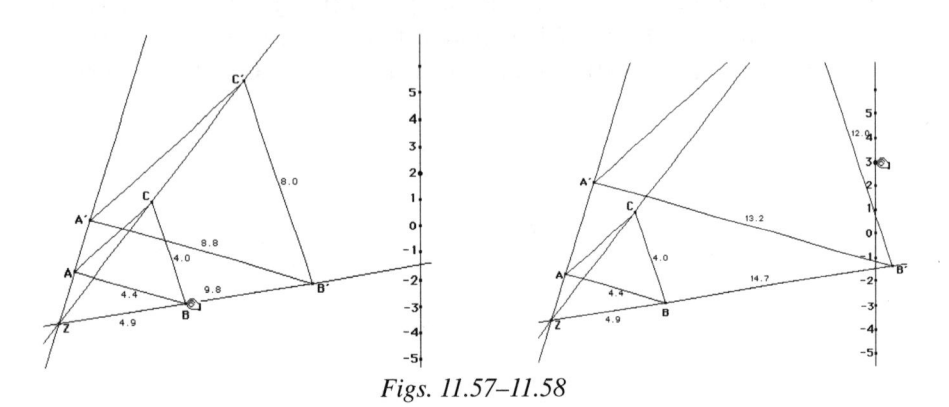

Example 17 *Touching circles with pre-given radius*
Problem: In Fig. 11.59, two circles with centres M_1, M_2 and radii r_1, r_2, are shown. Every possible circle with a given radius r which touches both circles is drawn in.

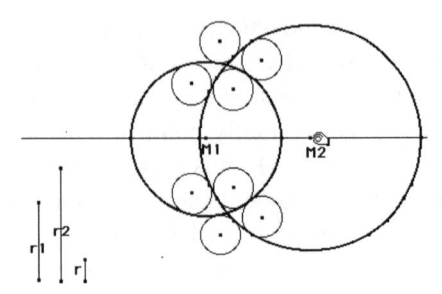

Figs. 11.59

Designing geometrical microworlds

a) Move M_1 or M_2 and investigate the number and position of the touching-circles of radius r (M_1 and M_2 should not coincide!)
b) Also vary r, r_1, r_2 and investigate whether there are differing numbers of touching-circles.
c) Which condition of r, r_1, r_2, M_1, M_2 results in *no* touching-circles?

Comment: The student should find distinct cases by systematically specialising a pre-given solution. All data (distance apart of the two midpoints, sizes of the three radii) can be varied.
In Figs. 11.60–11.63, M_2 is moved. A necessary and sufficient condition for the non-existence of touching-circles is: $|M_1 M_2| > 2 (r_1 + r_2 + r)$.

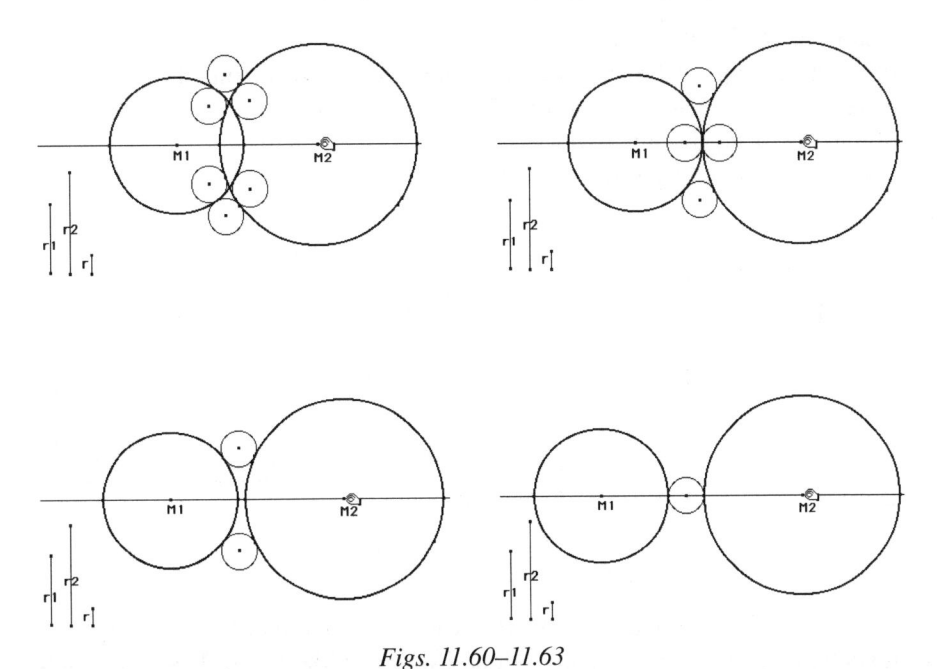

Figs. 11.60–11.63

The variation of r produces 3, 5, 7 touching-circles (Fig. 11.64 shows 5 touching-circles). In Fig. 11.65 we see a solution with the previously hidden help-lines revealed. The construction can be followed up in **History** mode. A question left to the reader is the general validity of the construction.

221

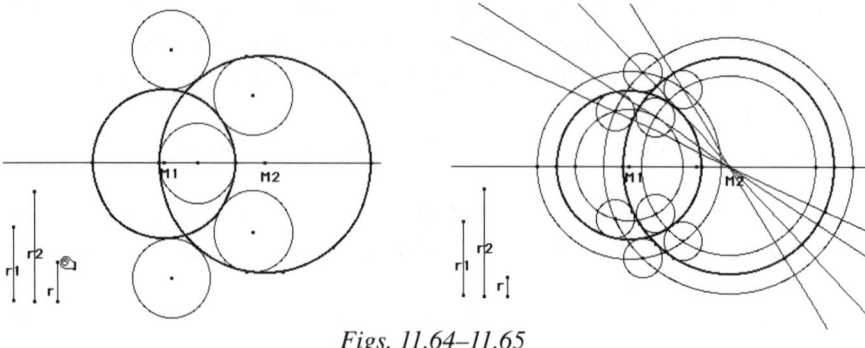

Figs. 11.64–11.65

Example 18 *Revolving table*
Problem: In Fig. 11.66, a revolving table-top is shown from above in both its revolved and original positions. Find the position of the pivot in the base of the table-top, about which it revolves.
Choose: **Locus of a point.**

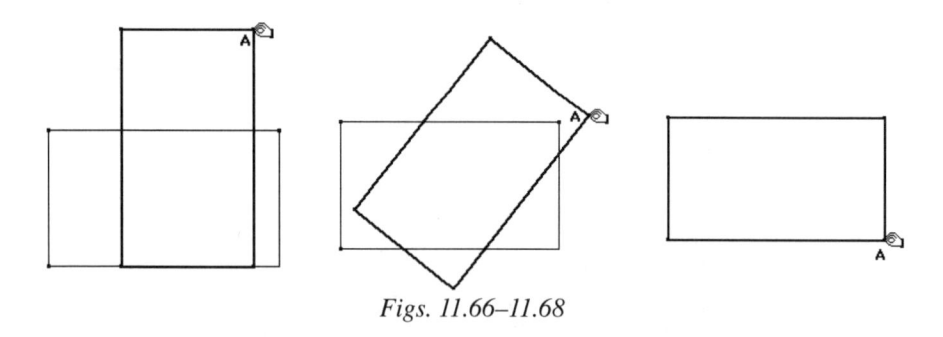

Figs. 11.66–11.68

Comment: This applied task in school level transformation geometry is a reversible problem. The menu options **Look of objects** and **History** which allow the user to see the underlying construction can be disabled to deny students such information, using **Edit menus**. In Figs. 11.67–11.68 we see the table in an intermediate position and then in the returned position. The tracks of the corners of the table, which are concentric circular arcs, are shown as loci in Fig. 11.69. The solution (Z is the pivot point) is obtained from an intersection of the perpendicular bisectors of two chords (Fig. 11.70).

Designing geometrical microworlds

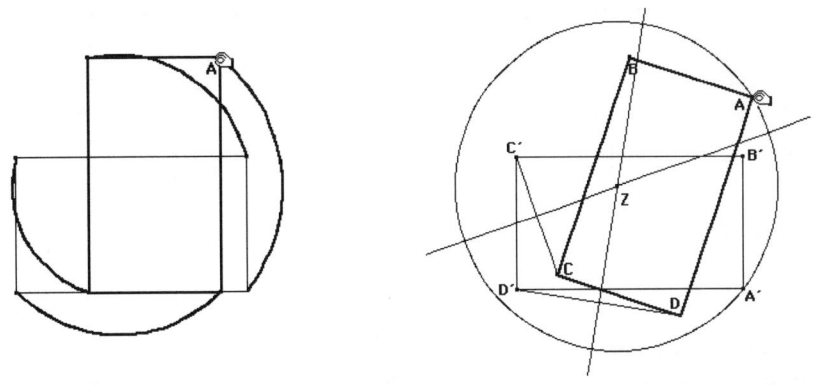

Figs. 11.69–11.70

Example 19 *Negotiating a corner*
Problem: The rectangle of dimensions 1.1 by 5.6 will not go round the corner (Fig. 11.71). For which dimensions can the rectangle be steered around the corner? (You can adjust the dimensions of the rectangle by dragging B and C, and the width of the corridor with D and E.)

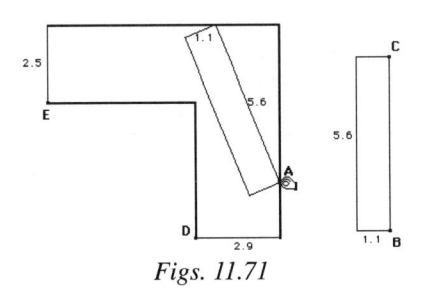

Figs. 11.71

Comment: The students have to recognise that there are two possible solutions: to make the rectangle either 'thinner' or to make it 'shorter'. The manipulation of the rectangle on the screen serves as preparation for the demanding exact solution of the task in which the dimensions of the rectangle and those of the corridor are included as variables. It can be seen in Fig. 11.72 that the rectangle does not fit around the corner; by shortening it you can succeed (Fig. 11.73). Maybe this will help to avoid unpleasant surprises when moving furniture!

223

Figs. 11.72–11.73

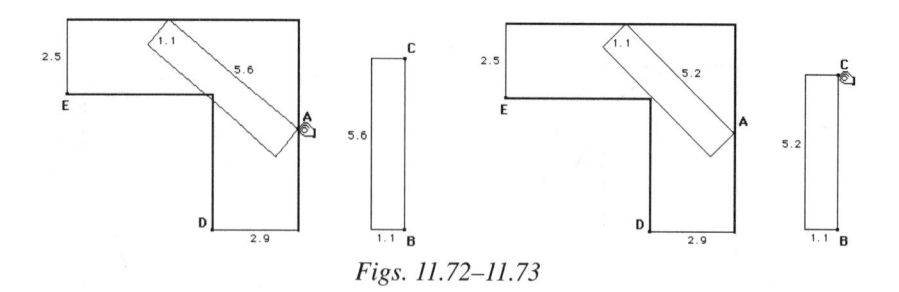

Example 20 *Trammel wheel*
Problem: A crank MA rotates around the point M; a rod is fixed at A which runs freely through C and whose end B describes a curve as the crank revolves (Fig. 11.74). Find out for which crank length (adjustable by moving D, with |MD|=|DC|) and for which rod length (adjustable by moving B) a major part of the curve followed by B is straight and perpendicular to MC.
Choose: **Locus of a point**.

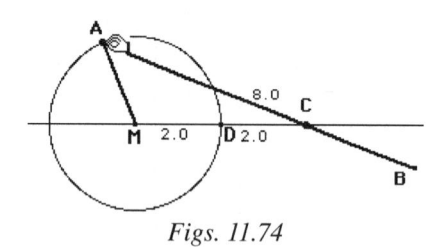

Figs. 11.74

Comment: Mechanisms which transfer a circular motion exactly into a rectilinear movement such as the invertors of Peaucellier and of Hart, have the disadvantage of consisting of a relatively large number of joints and linkages. For reasons of technical safety such gears are less suitable. An *approximate* straight-line guidance is often sufficient which can be accomplished by numerous other jointed mechanisms which have fewer joints or linkages. Among them is the trammel wheel. The name comes from the constraining of the rod to pass through C (Fig. 11.74). With the given values the student obtains a curve such as in Fig. 11.75, which does not show the desired linearity. Only with adjustment, such as indicated in Fig. 11.76, is a suitable solution obtained.

Able students aged 16+ can analytically describe the curve in Fig. 11.76. To do this, it is best if C is selected as the origin of the coordinate system and the angle of rotation of the crank (relative to the x-axis) is introduced as a parameter to be eliminated. This results in an algebraic curve of degree 6, implicitly represented by $(x^2 + y^2)(x^2+y^2+4x+84)^2 - 324(x^2+y^2+2x)^2 = 0$.

The straight line x = 6 fits closely to the linear part of this curve.

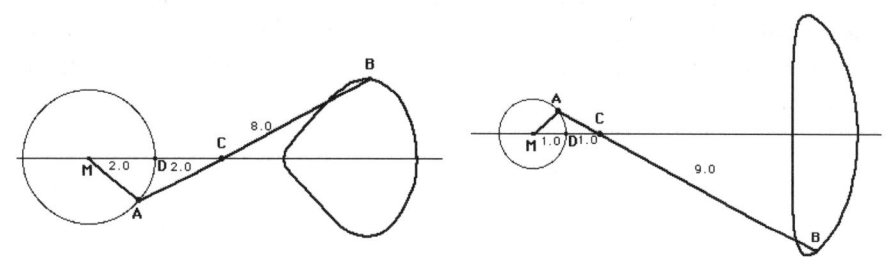

Figs. 11.75–11.76

Example 21 *Reciprocating pump*
The driving wheel rotates about M; rod AB pushes the piston in the cylinder back and forth (Fig. 11.77).

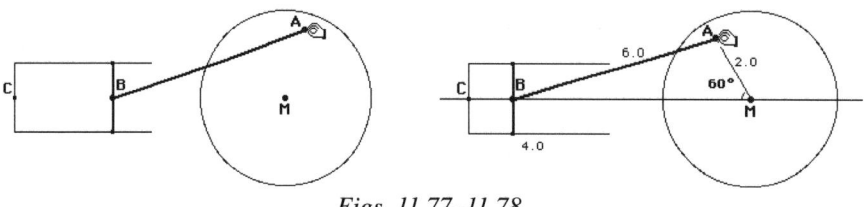

Figs. 11.77–11.78

(1) (a) When does the piston's direction of movement reverse?
   (b) On which quantity does the length of the cylinder depend?
   (c) Let angle BMA be 60°, the crank length MA 2.0 units and the rod length AB 6.0 units.
   What is the distance BC?
   (Add to the construction as necessary.)
   (d) If the remaining piston stroke (BC) is 2.0 units what is angle BMA?
   (e) What is the piston capacity? (Assume the cylinder's cross-sectional area is 1 sq. unit.)

Discovering Geometry with a Computer

(2) (a) Find out how this simple model was constructed.
   (Activate all the concealed auxiliary lines and points, modify the size of the pushing rod AB and that of the crank AM.)
   (b) Construct such a model on your own. (Use the three macros COPY-LENGTH, CIRCLE-C-r and SHAFT.)
   (c) Test the functional efficiency of your model.

Fig. 11.78 shows a schematic model. In Fig. 11.79 the construction is shown with auxiliary circles, lines and points (the determining parameters are beneath the figure and represent the lengths of the connecting rod and of the crank).

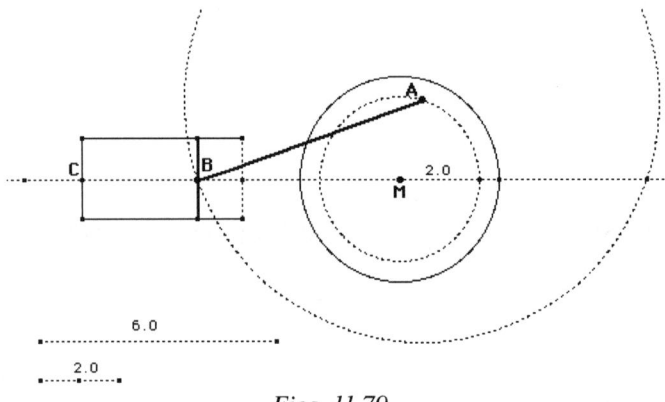

Figs. 11.79

11.5 MODELLING NON-EUCLIDIAN GEOMETRIES – AN ADVANCED TOPIC

A major purpose of Cabri-Géomètre microworlds is the modelling of plane geometries with compasses and ruler for a better understanding of the geometry of the Euclidean plane – as developed axiomatically in David Hilbert's monumental work of 1899 *Fundamentals of Geometry*.

The set of basic geometric elements of non-Euclidean geometries can be obtained by restriction or modification of the basic Euclidean objects and properties. In the following we restrict ourselves to one model of the non-Euclidean geometry known as hyperbolic geometry.

Designing geometrical microworlds

Example 22 *Klein's model of hyperbolic geometry*

Klein's model of hyperbolic geometry can be developed via polar reflection (h-line reflection). The h-points are the points in the interior of a circle; the h-lines are chords of the circle without endpoints. Fig. 11.80 shows an appropriate **Construction** menu.

```
Construction
 Locus of point
 Point on object
 Intersection
 ......................................
 h-line
 h-border parallels
 h-line reflection
 h-orthogonal
 h-midpoint
 h-mid-perpendicular
 h-point reflection
 h-angle-bisector
 h-angle measurement
```

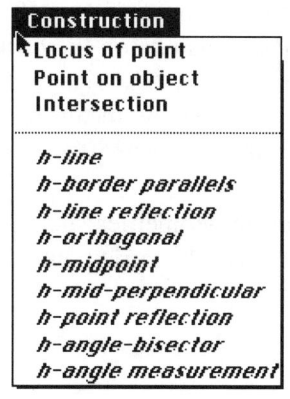

Figs. 11.80

One of the main characteristics of hyperbolic geometry compared to Euclidean geometry is the existence of several (non-intersecting) parallels p through an h-point P outside an h-line g. The macro ***h-border parallels*** produces the so-called limiting parallels p_1, p_2 (Fig. 11.81); none of the h-lines p, which span the corresponding angle fields fixed by the limiting parallels and which pass through P, intersect g.

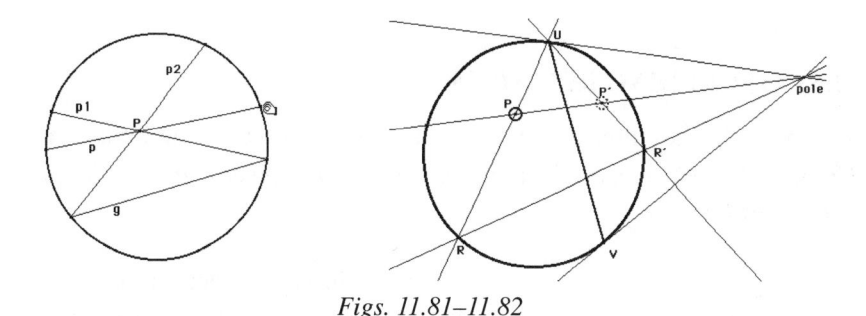

Figs. 11.81–11.82

The ***h-line-reflection*** is based on the so-called polar reflection (macro definition illustrated in Fig. 11.82). With the h-line and boundary points U, V given; we construct the pole of the polars U, V. The image point P' for a

227

given h-point P after reflection in UV is constructed by means of the properties of fixed points and fixed straight lines analogous to Euclidean reflection in straight lines. For the definition of the macro, the circle, the segment line UV and point P are initial objects and point P' is the target object. All further h-commands in the construction menu use *h-line-reflection* and the concepts analogous to Euclidean reflection geometry. Fig. 11.83 illustrates the macro definition of the orthogonal through P to the h-line with the peripheral points, U, V (P can also be on the h-line). Fig. 11.84 shows the measurements of the interior angles of a triangle, which can be measured in the Euclidean way as centre angles by means of *h-angular-measurement*; the angular sum is always smaller than 180°.

A treatment of non-Euclidean geometries such as Klein's model of hyperbolic geometry can be successfully carried out with able students aged 16+ working in groups.

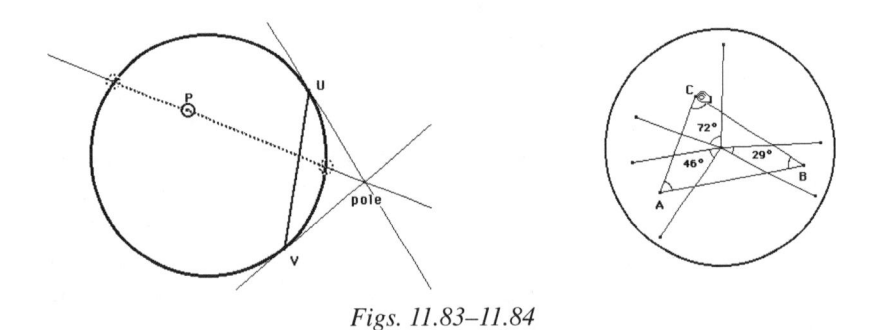

Figs. 11.83–11.84

11.6 CONCLUDING REMARKS

We conclude with three comments, which we regard as being important.

Comment 1 *(A technical limitation of Cabri-géomètre in the design of microworlds)*

Certain general macros cannot be defined through screen drawings. This can be seen from the example of the macro for drawing the two common external tangents to two circles (macro EXTERNAL-TANGENTS-TO-CIRCLES). Although the on-line help (Fig. 11.85) is followed (resulting in Fig. 11.86), the construction proves to be faulty if the smaller circle is changed into the bigger circle using drag-mode (holding down the OPTION key so that the basic circle can be re-sized) (Fig. 11.87). The problem can only be overcome by coding in a geometric programming language with appropriate control structures. This has yet to be implemented in Cabri-géomètre.

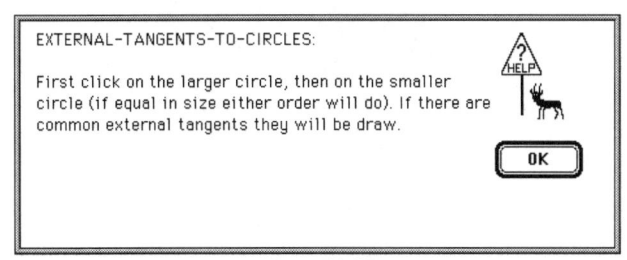

Figs. 11.85

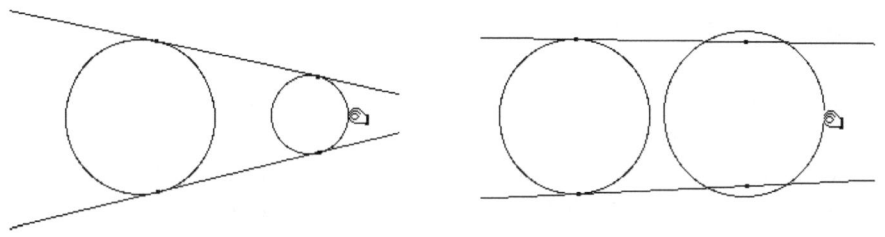

Figs. 11.86–11.87

Comment 2 *(A developmental perspective of a plane geometry tool)*
When designing microworlds it is desirable to have a flexible, interactive and modular tool which allows the processing of all objects of plane geometry useful in mathematics education. Compatible system modules, illustrated in Diagram 11.1, could be:

(a) a construction system (optionally menu driven or command driven);
(b) a calculation system (a numerical module compatible with the drag mode of the construction system);
(c) an algebra system (a symbol processing module with an interface to the calculation system which allows generalisations);
(d) a hypothesis testing system (a 'mechanical geometry theorem proving' facility);
(e) a programming system (a friendly graphic user language for direct manipulation of screen objects) with printing of figures available.

Discovering Geometry with a Computer

```
                    Geometric
                    Constructic

     Programming                    Calculati


         Hypothesis test:    Algebraic manipulat
```
Diagram 11.1

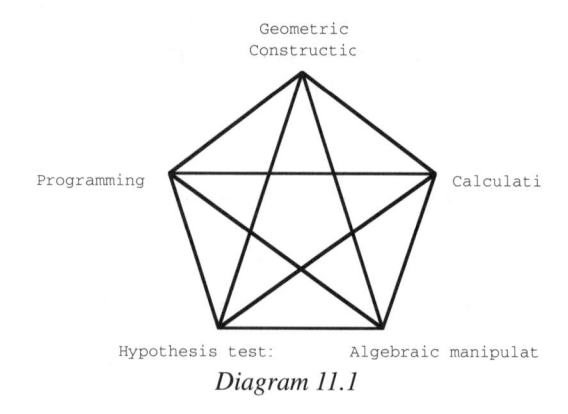

It is evident that the creation of such a system, which permits the universal design of geometric microworlds, will require a considerable effort by a team of geometers, mathematics educators, information technologists, programmers and teachers of geometry!

Comment 3 *(Considerations for the use of educational software for learning in microworlds)*
The use of educational software allows only certain teaching functions and learning processes to be optimised, and not teaching and learning as complete entities. (One only needs to think of the distinction between primary and secondary experience!) Teaching media and methods and therefore teaching programmes must complement each other.

The learning and teaching process as a whole is essentially dependent on forms of personal interaction and communication. The use of educational software, as an impersonal form of communication, must therefore be embedded in personal forms of teaching. The use of mainly closed educational software presupposes, as a rule, the acceptance of educational decisions about aims, content and methods. This may restrict the scope for decisions and approaches of teachers and students, and conceal the threat of the curriculum being controlled from outside.

The development of their own educational software by teachers and students is very time-consuming and the results rarely meet acceptable software-ergonomic standards. The (expensive) development of educational software is the task of professional teams of experts. Extensive trialling is absolutely essential. The teaching situation is gradually changing due to the use of educational software and teachers are facing new challenges. They must be put in a position to be able to integrate suitable educational software into teaching in an effective and fair way and this demands appropriate initial and in-service teacher training.

CHAPTER 12

Microworlds for isoperimetric problems

'Tis distance lends enchantment to the view.
Thomas Campbell

12.1 INTRODUCTION

An isoperimetric problem in the Euclidean plane is usually expressible as:

"In a class of figures with equal perimeters, is there one or more with a maximum area?"

Assuming such figures with maximum area exist, the question is to specify the solution as clearly as possible. The problem of existence is generally the more difficult problem. The same applies to the dual to the isoperimetric problem, the iso-areal problem:

"In a class of figures with equal areas, is there one or more with a minimum perimeter?"

The existence and clarity of the solution of an iso-areal problem can be directly linked to that of the corresponding isoperimetric problem, and vice-versa. We can limit ourselves to figures which are convex, because for every non-convex figure with a given finite perimeter (or area), a corresponding convex figure with larger area (or smaller perimeter) can be constructed.

In lower secondary school geometry teaching, polygons (with a fixed number of sides) are suitable subjects for experimental isoperimetric problems. A polygon (n > 3) is here understood to be freely jointed at all its vertices. Isoperimetric and iso-areal problems can help the student to learn how to abstract, generalise, specialise, make analogies, etc.

Microwolds for isoperimetric problems

The following working objectives may be adopted in geometrical experiments by pupils:

1(a) The student should observe that varying the shape of a figure while holding its perimeter constant generally alters the size of its area.

1(b) The student should observe that varying the shape of a figure while holding its area constant generally alters the size of its perimeter.

(These two facts are by no means obvious to every student.)

2(a) The student should find out which of a particular group of isoperimetric figures (e.g. all triangles with perimeter 3 units) has the largest area.

2(b) The student should find out which of a particular group of iso-areal figures (e.g. all rectangles with area 4 sq. units) has the smallest perimeter.

The concepts which relate the perimeter and area of polygons are important, although in most geometry teaching the area and perimeter of a polygon are not treated as being functionally connected and arithmetical-algebraic aspects are emphasised. Perimeter-invariant operations on figures do not appear at all in the syllabuses.

Computer microworlds can be set up in systems such as Cabri-géomètre to make the above objectives more easily attainable when used in combination with the conventional approaches. With such a graphics system, the user can manipulate geometrical objects on screen to investigate perimeter-invariant (and area-invariant) operations on polygons.

We now describe in general terms such an experimental microworld for a standard isoperimetric problem (presented in Example 1, Section 12.2):

The student loads a prepared configuration from disc into the graphics system, and at the top of the screen finds the given perimeter represented as a single line segment (which we will call the 'perimeter-line') together with its measurement (see Fig. 12.1 for an example). The perimeter-line is divided into n variable subdivisions. Out of these subdivisions, an n-gon is constructed and displayed in the middle of the screen. At the bottom of the screen, the area of the n-gon is represented as a rectangular strip with height 1 unit together with the measurement of its length which represents its area. The student now has the following possibilities.

The student can cycle through the following actions:

(1) Change the subdivisions of the perimeter line by dragging with the graphics cursor and thus vary the shape of the n-gon;
(2) Change the shape of the polygon (altering the subdivisions of the perimeter-line by dragging vertices of the (freely jointed) polygon;
(3) Change the size of the perimeter by dragging an extreme point of the perimeter-line.

The student needs no particular competence in graphic construction nor familiarity with the system in order to work in such a problem-specific microworld. However, since the microworld is open to the full Cabri-géomètre graphics system, the student *could* carry out additional constructions and measurements, e.g. generate loci, draw circles, mark and measure angles, etc. The curious student could make visible the hidden auxiliary lines, which are used for the construction.

Every experimental environment requires:

- At the start: a statement of the problem and brief instructions for the use of the interface.
- During the experimenting: the advice of the competent teacher (this is essential).
- At the end: a written report by the student. The final solutions reached could be captured by a screen dump which the student could annotate.

The solutions found in experimental microworlds do not necessarily – encourage deep reasoning. According to the student's ability and age, it is helpful to follow up such work with ruler-and-compasses constructions (both with and without the aid of the computer), as well as numerical treatment and the use of mathematical formulas or special apparatus. Other computer based work (e.g. function plotting) is of course possible but the lack of standardisation raises significant problems when several different systems are used.

12.2 ISOPERIMETRIC PROBLEMS INVOLVING QUADRILATERALS

Our choice here mainly reflects the wish to emphasise the potential for relatively early use (say by 13/14 years olds). To understand the tasks such students will need some instruction. The following examples are not arranged in order of increasing difficulty.

Example 1 *Equi-perimetric quadrilaterals*
We vary the side-lengths of the quadrilateral which has a constant perimeter of magnitude 12 units (Fig. 12.1) until two opposite sides are equally long (Fig. 12.2); the area has become smaller. Now we make the quadrilateral into an isosceles trapezium by dragging one vertex (Fig. 12.3). Starting with Fig. 12.1 again, we draw a circle through three vertices by using a supplied macro 3-POINT-CIRCLE (Fig. 12.4). The circle and quadrilateral can be changed by dragging the remaining vertex until it comes to lie on the circle (Fig. 12.5); the area of the cyclic quadrilateral is larger than the area of any other quadrilateral with sides of the same lengths. If we change the quadrilateral into a kite (Fig. 12.6), the kite with two opposite right-angles has the largest area among the set of kites with sides of the same lengths (Fig. 12.7). If we make a rectangle with sides the same lengths as those of this kite, the area does not change (Fig. 12.8). If three sides are made equal, then *all* are equal and a rhombus results (Fig. 12.9), whose moveable vertices travel on congruent circles, as can be seen with the help of the **Locus** command (Fig. 12.10). The rhombus' area varies through all values from zero up to the largest, which occurs when the rhombus becomes a square (Fig. 12.11). It is clear that its area is not exceeded by any other quadrilateral with the same perimeter. We now vary the magnitude of the perimeter and confirm our conjecture experimentally (Fig. 12.12).

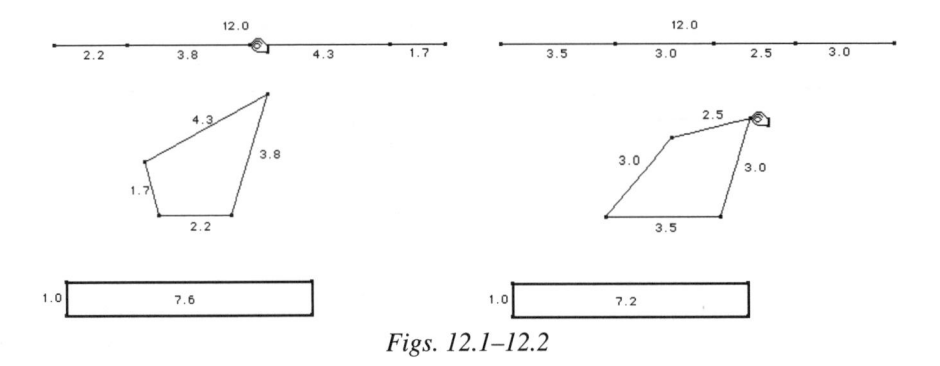

Figs. 12.1–12.2

Discovering Geometry with a Computer

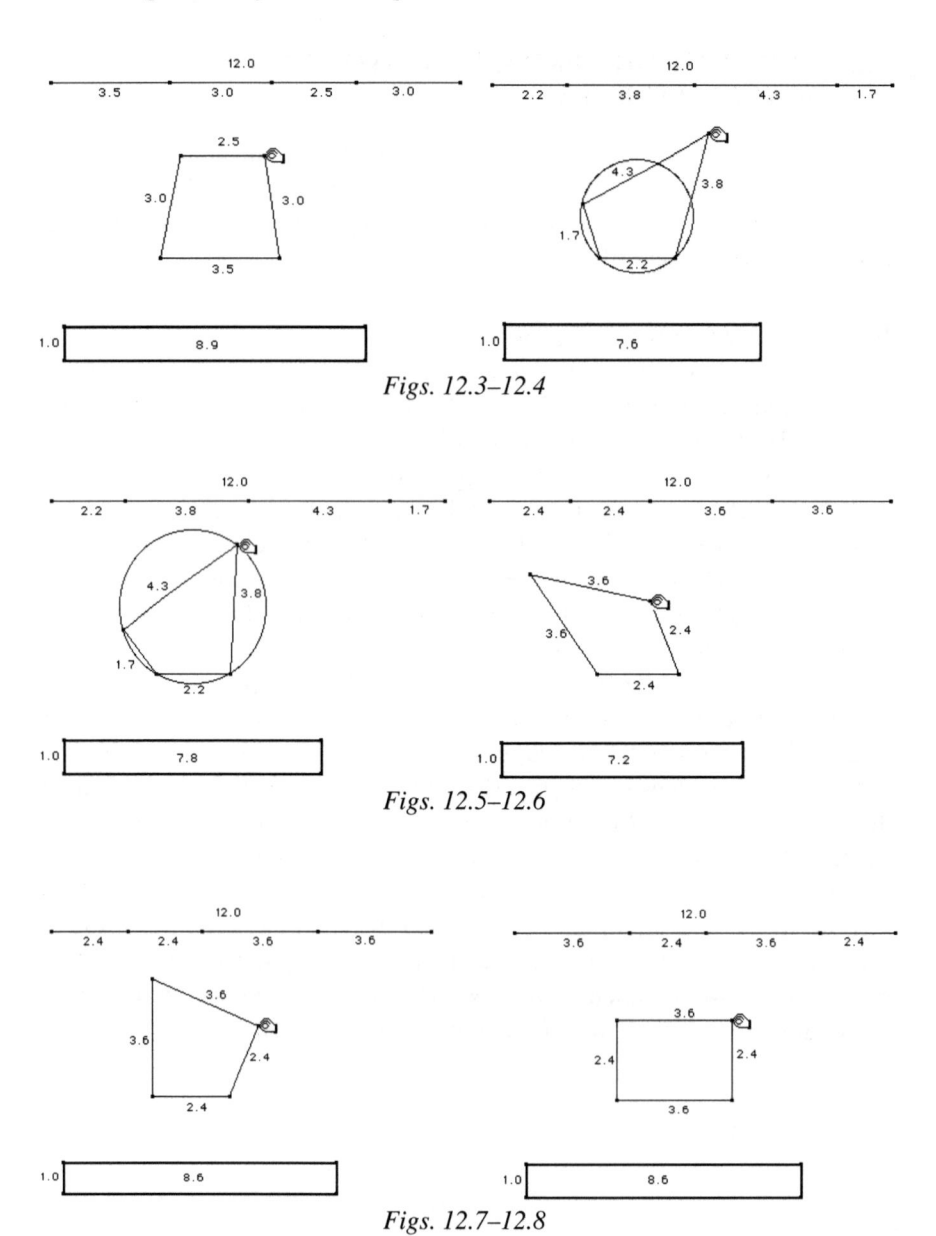

Figs. 12.3–12.4

Figs. 12.5–12.6

Figs. 12.7–12.8

236

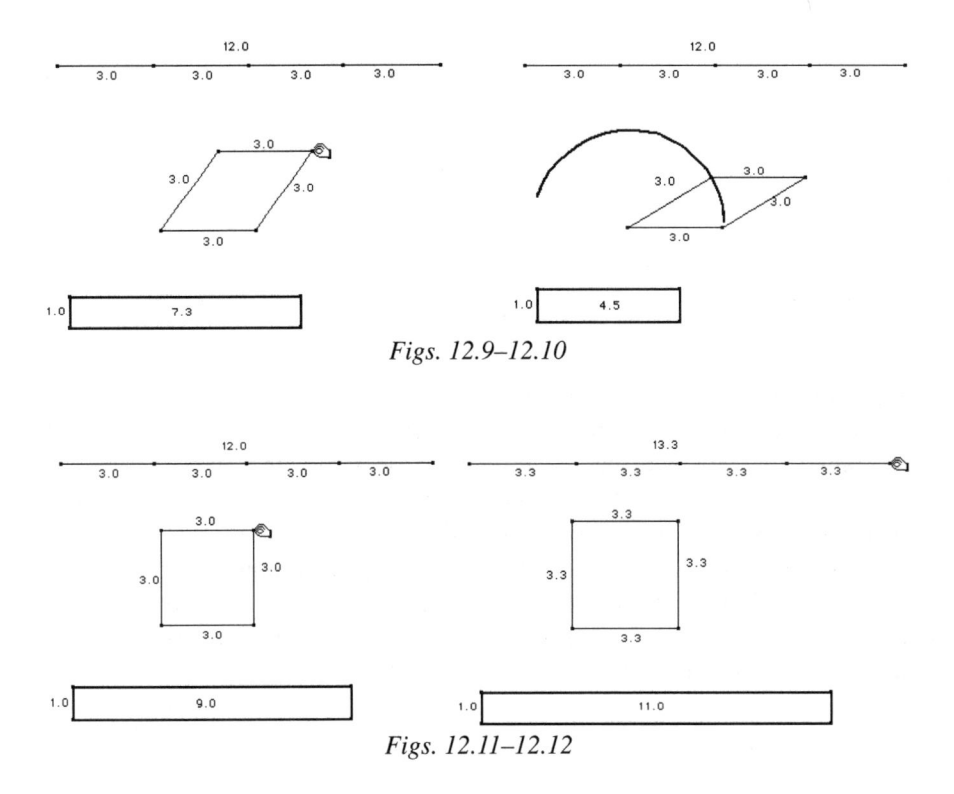

Microwolds for isoperimetric problems

Figs. 12.9–12.10

Figs. 12.11–12.12

Example 2a *Equi-perimetric rectangles*
We vary the shape of a constant-perimeter rectangle (Fig. 12.13; given perimeter 11.6 units) and recognise the linear connection between the side-lengths with the help of the **Locus** command (Fig. 12.14). The area is at its largest when the rectangle becomes a square (Fig. 12.15). We now vary the magnitude of the perimeter and confirm our conjecture experimentally (Fig. 12.16).

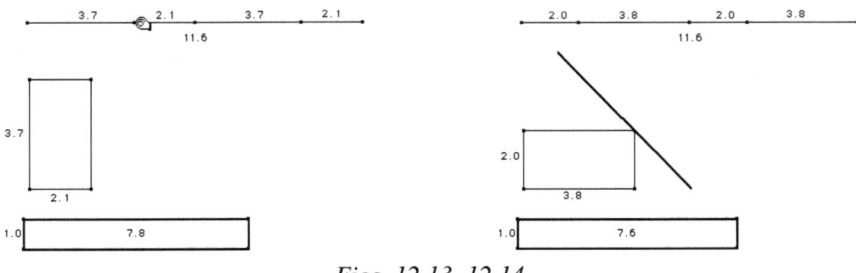

Figs. 12.13–12.14

237

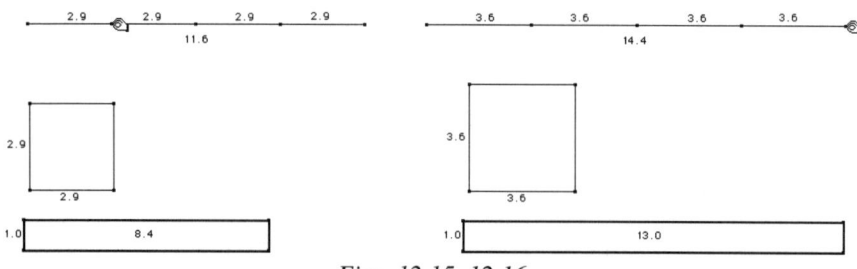

Figs. 12.15–12.16

Example 2b *Equi-areal rectangles*
We vary the shape of a constant-area rectangle (Fig. 12.17; given area 4.0 sq. units) and observe the magnitude of the perimeter: it decreases (or increases) when the difference between the rectangle sides decreases (or increases). We recognise the functional (inversely proportional) relationship between the side-lengths with the help of the **Locus** command (Fig. 12.18; hyperbola). The perimeter becomes minimal when the rectangle becomes a square (Fig. 12.19). Variation of the given fixed area (Fig. 12.20) leads to experimental verification of our result.

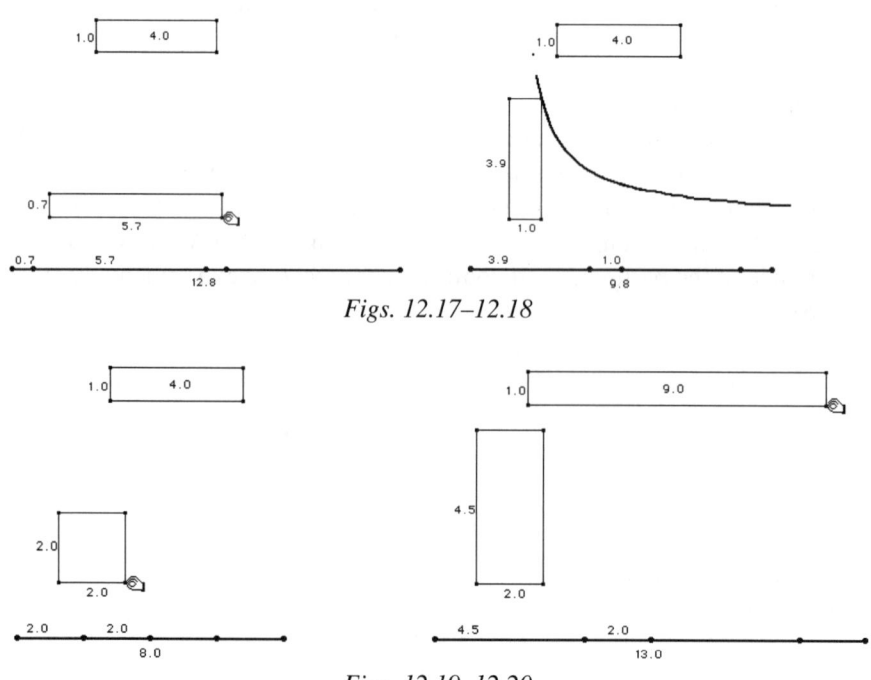

Figs. 12.17–12.18

Figs. 12.19–12.20

Microwolds for isoperimetric problems

Example 3 *Equi-perimetric isosceles trapezia with a fixed base*
We vary the lengths of the variable sides of an isosceles trapezium with a given base and explore the limiting positions (Figs. 12.21–12.22; the interested reader can determine the condition for the existence of such a trapezium – in the case of non-existence, the graphics system erases the figure, which reappears when it is changed back to a feasible figure). As a special case, we can obtain a rectangle (Fig. 12.23). During variation of the side-length, the moveable vertex of the trapezium moves on a parabolic path, which we can see using the **Locus** command (Fig. 12.24). Physical apparatus to perform this is shown in Fig. 12.25 (l = guide-line, F = focus). The area is maximal when the three adjustable sides are equal (Fig. 12.26).

A question for the reader: what is the corresponding iso-areal problem and its solution?

Figs. 12.21–12.22

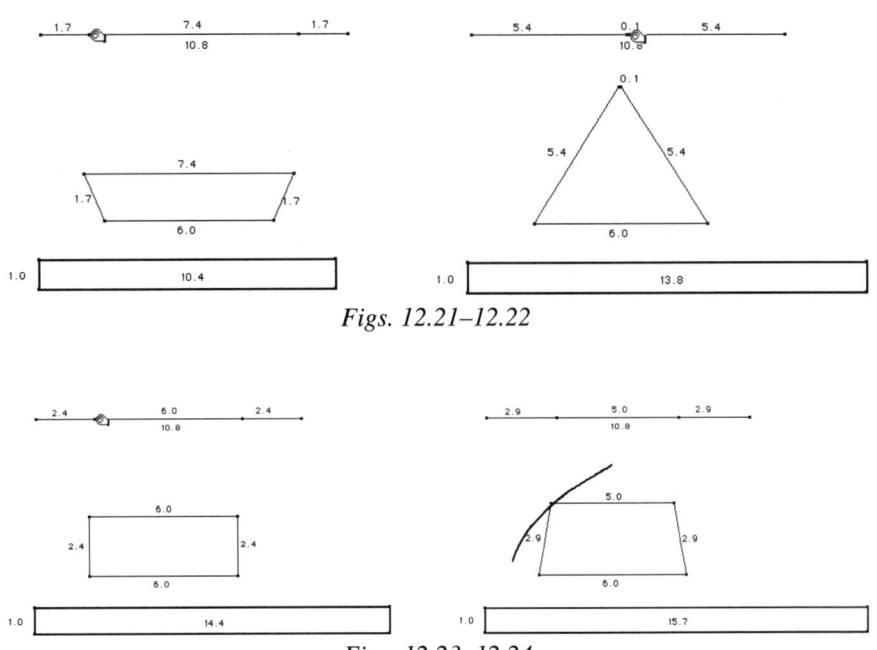

Figs. 12.23–12.24

Figs. 12.25–12.26

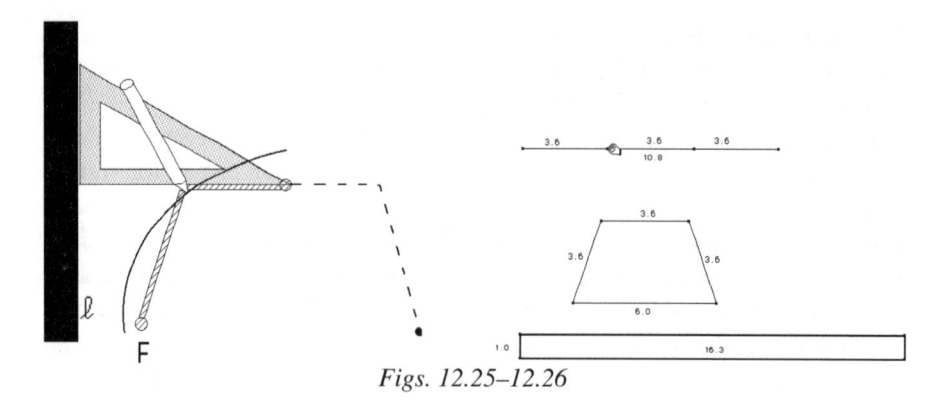

Example 4 *A real world isoperimetric task*
Mr Capon has a garden, which is bordered by a long wall. He wants to fence off a rectangular piece of his garden for his hens using 10.8m of wire netting. In so doing, he uses the wall as one of the sides of the rectangle. How long and how wide must the rectangle be, so that the enclosure is the largest possible?

An experimental (trial) arrangement is shown in Fig. 12.27. We make the enclosure square (Fig. 12.28), but find that the area has been reduced. We obtain the maximum area when the longer side runs parallel to the wall and is twice as long as the shorter sides (Fig. 12.29).

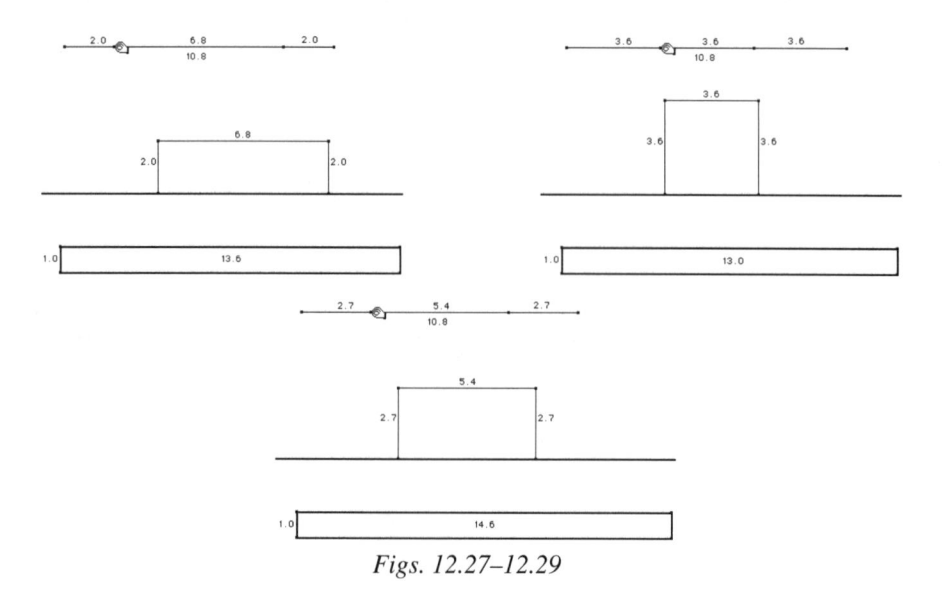

Figs. 12.27–12.29

Such a piece of garden is to be fenced off for 8 (or 9) hens for whom a running area of one square metre per hen is required. What measurements should the enclosure have if as little wire netting as possible is to be used? Experimental solutions of the geometrical model for this problem are shown in Fig. 12.30 for 8 hens and in Fig. 12.31 for 9 hens.).

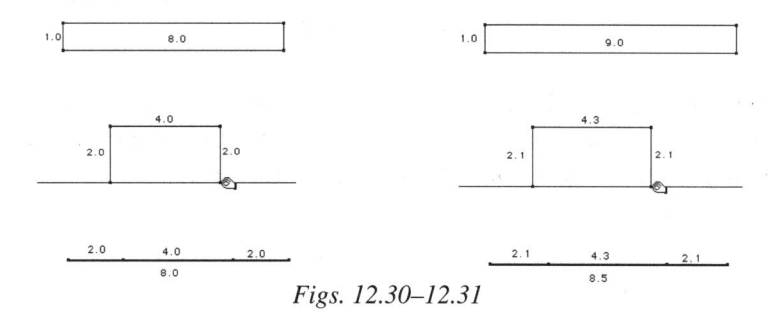

Figs. 12.30–12.31

12.3 ISOPERIMETRIC PROBLEM INVOLVING POLYGONS

The automatic calculation of polygonal area (available in later versions of Cabri-géomètre) is achieved by clicking on the vertices around the area (for up to 10-sided figures). This allows us to find experimental solutions for a range of shapes – here illustrated for the pentagon.

In Fig. 12.32 we vary the side lengths of a pentagon and measure the respective area: Area ABCDE = 14.6. As we make all sides equally long and the pentagon more regular to the eye (Fig. 12.33), the area becomes larger. We now mark the interior angles (Fig. 12.34), measure them and make the pentagon as regular as possible (Fig. 12.35). The area is maximal and found to be 16.5 sq. units (for perimeter 15.5 units), which tallies with the calculated numeric value, rounded to one decimal place.

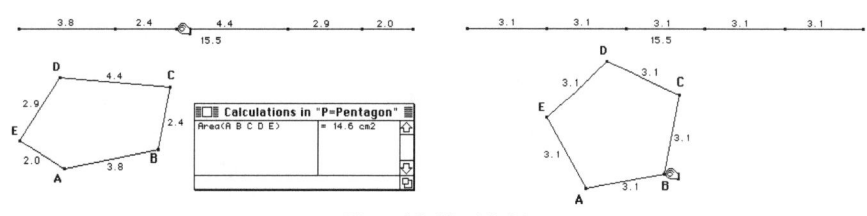

Figs. 12.32–12.33

241

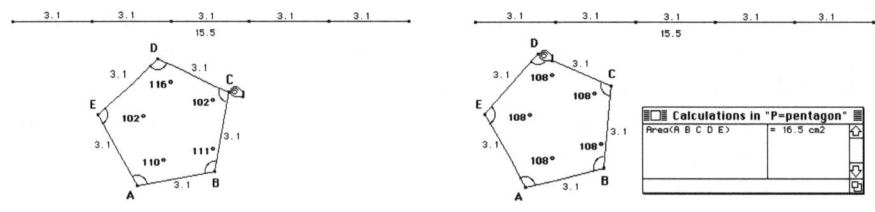

Figs. 12.36–12.37

We could alternatively have found the optimal pentagon by passing a circle through three corners (Fig. 12.36) and varying the pentagon until the remaining corners also lay equally spaced round the circumference of the circle (Fig. 12.37).

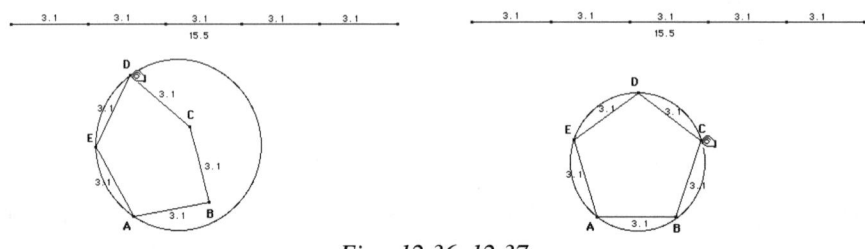

Figs. 12.36–12.37

12.4 FINAL REMARKS

The above approach only becomes fully effective with adequate hardware and software (including prepared configurations). Ideally about two students per computer should be provided but even with just one system connected to an OHP at the front of the class some useful work is possible.

CHAPTER 13

Microworlds for geometrical transformations

> *The universe is transformation.*
> Marcus Aurelius Antoninus

13.1 INTRODUCTION

This chapter contains an introduction to the well-known transformations:

– translation
– line reflection
– rotation
– enlargement.

These are documented by phase diagrams with very little discussion, as the material itself will be familiar to the reader. Only for translation is the determination of the fixed lines or fixed circles made explicit.

Microworlds for combining transformations can be designed by defining appropriate macros. The method of creating the required macros is not given here. This may be treated as an investigation for the teacher, or the macros supplied on the disc associated with this book can be analysed or simply used. It is recommended that written instructions for each investigation are prepared for the students, and prepared figures supplied for loading with Cabri-géomètre.

13.2 FREE MOVEMENT OF A TRIANGLE

Three line segments are copied to form a triangle ABC. The free movement of the triangle (without deformation) can be produced using **Circle by centre & rad. point** and by selecting **Point on object** to attach B and C to two concentric circles centred on A (Fig. 13.1, showing auxiliary circles which would normally be hidden).

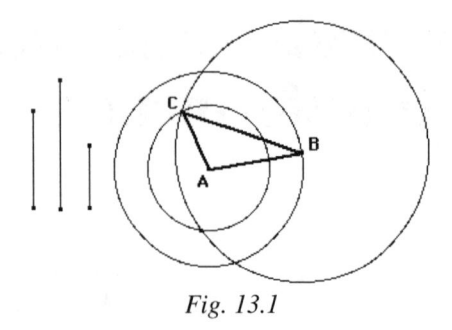

Fig. 13.1

Dragging basic point A allows the *translation* of the triangle (Figs. 13.2–13.3). Dragging basic point B, allows the *rotation* of the triangle about A (Fig. 13.4).

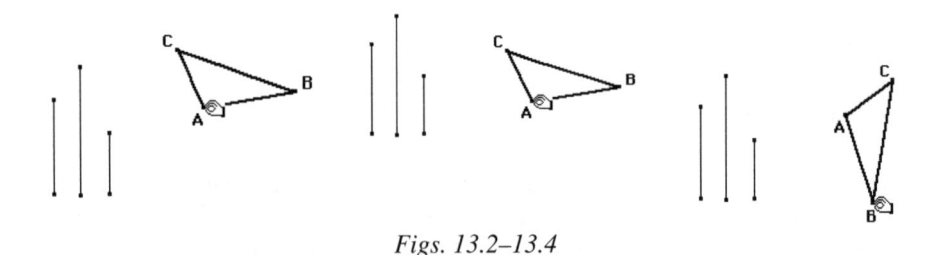

Figs. 13.2–13.4

13.3 TRANSLATION

Fig. 13.5 shows the fixed line which determines the direction of the translation. The magnitude of the translation is determined by a variable line segment lying on the line. Fig. 13.6 shows the paths of the image points, using copies of the line segment determining the translation.

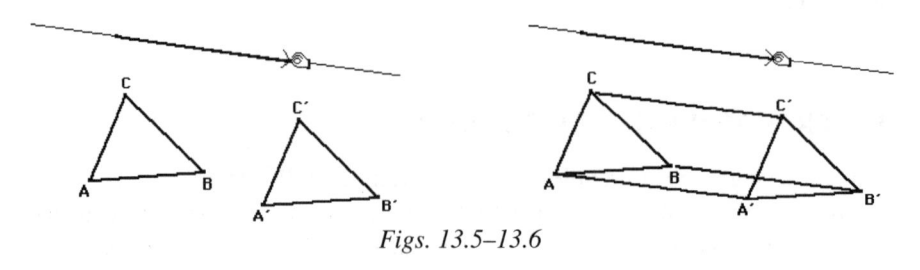

Figs. 13.5–13.6

Fig. 13.7 shows variation of the determining line segment (changing both length and direction from Fig. 13.5) and Fig. 13.8 shows a second variation.

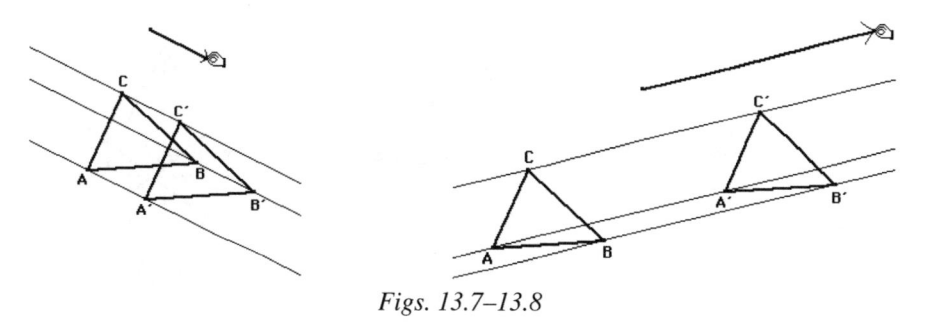

Figs. 13.7–13.8

Figs. 13.9–13.10 show the effect of varying the triangular figure by dragging C.

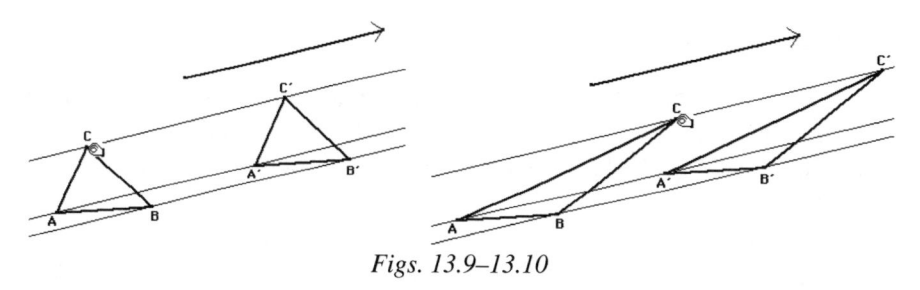

Figs. 13.9–13.10

13.4 LINE REFLECTION

Figs. 13.11–13.12 show the effect on the image of triangle ABC by changing the position of the axis of reflection.

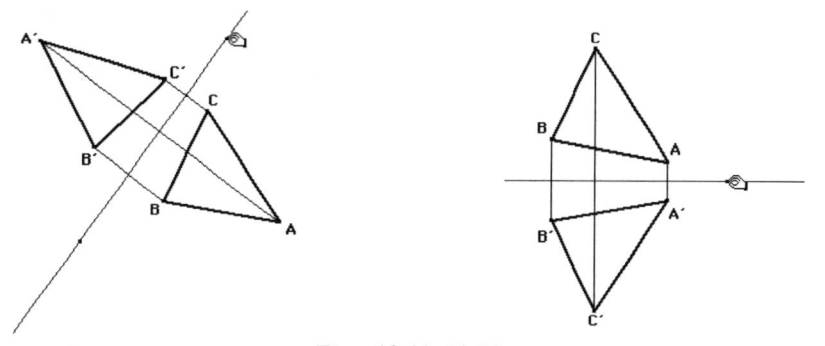

Figs. 13.11–13.12

Discovering Geometry with a Computer

Figs. 13.13–13.15 show the effect of varying the triangular figure by dragging A.

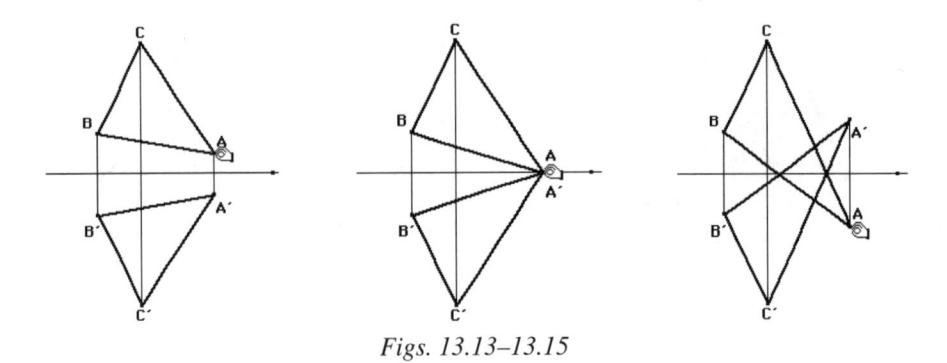

Figs. 13.13–13.15

13.5 ROTATION

Fig. 13.16 shows an initial position with the angle of rotation (59°) about to be varied and Fig. 13.17 shows the result of for the special case of a half-turn (180°) and Fig. 13.18 for a different angle of rotation (132°).

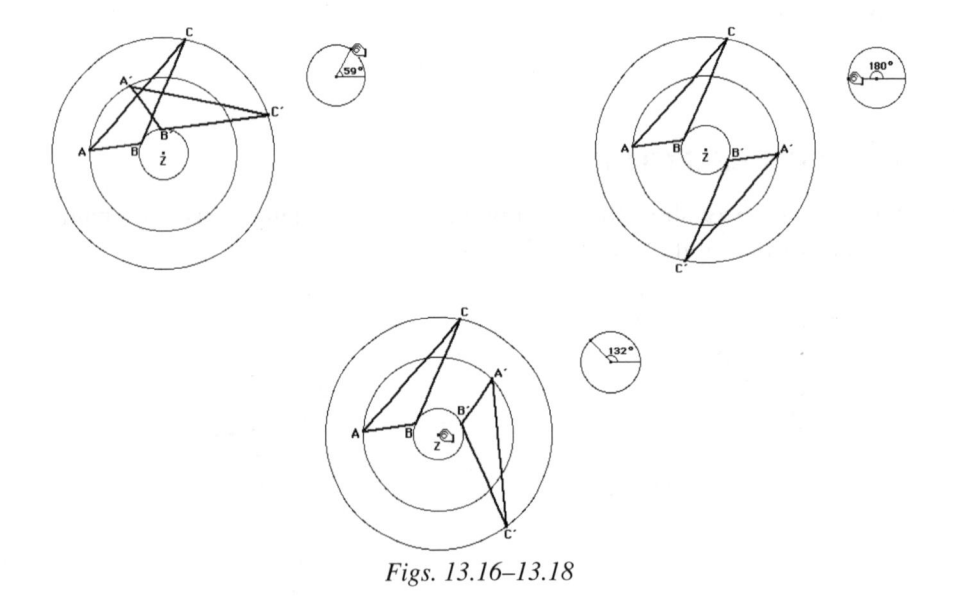

Figs. 13.16–13.18

Figs. 13.19–13.20 show the results of varying the position of the centre of rotation (Z) starting from Fig. 13.18.

Figs. 13.19–13.20

Figs. 13.21–13.22 show the effect of varying the triangular figure by dragging C.

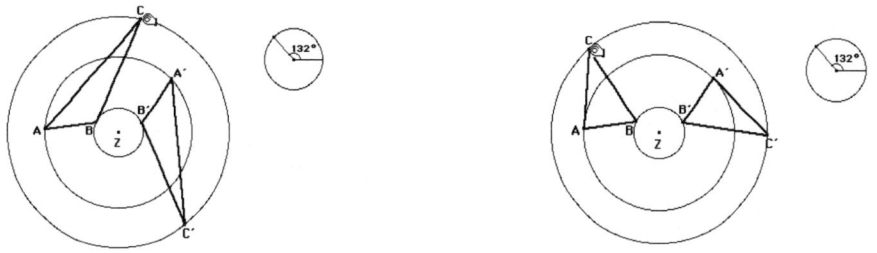

Figs. 13.21–13.22

13.6 ENLARGEMENT

Fig. 13.23 shows an initial position with scale factor set on the sliding scale to 2.5 and Fig. 13.24 shows the result of changing the scale factor to –1.5.

Discovering Geometry with a Computer

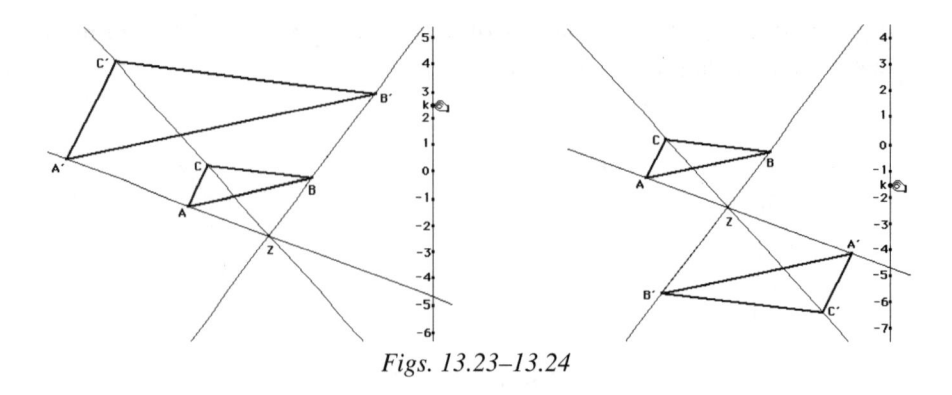

Figs. 13.23–13.24

Figs. 13.25–13.26 show the effect of varying the centre of enlargement.

Figs. 13.25–13.26

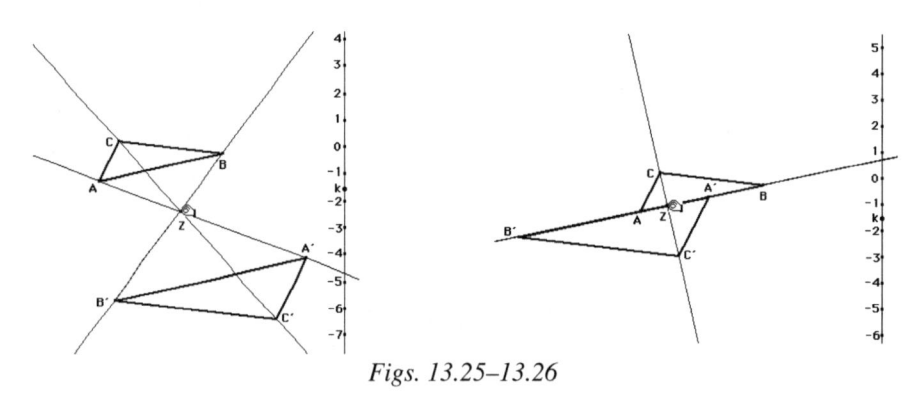

Figs. 13.27–13.28 show the effect of varying the triangular figure by dragging C.

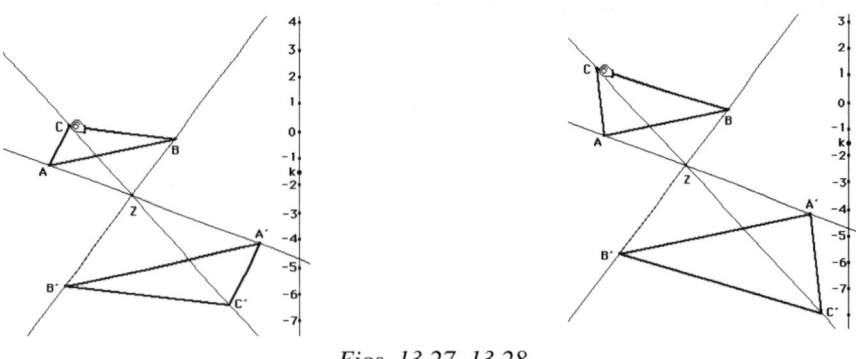

Figs. 13.27–13.28

248

Fig. 13.29 shows the effect of setting the scale factor to −1 (specialisation resulting in point reflection).

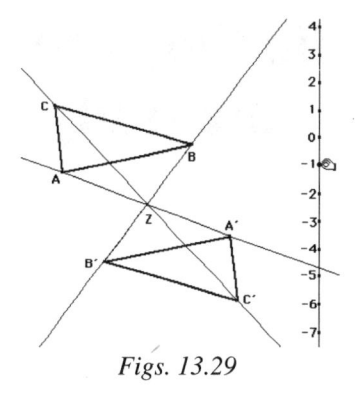

Figs. 13.29

In Figs. 13.30–13.31, O is the centre of enlargement and the scale factor is given by |OP'|/|OP|. In drag mode this ratio is invariant so we can use it to generate a scaled copy of any freehand drawn figure by moving P. To do this choose the **Locus** command and while holding down the SHIFT key click on the points whose loci one wants (in this case P' and P), and then drag the independent point (in this case P).

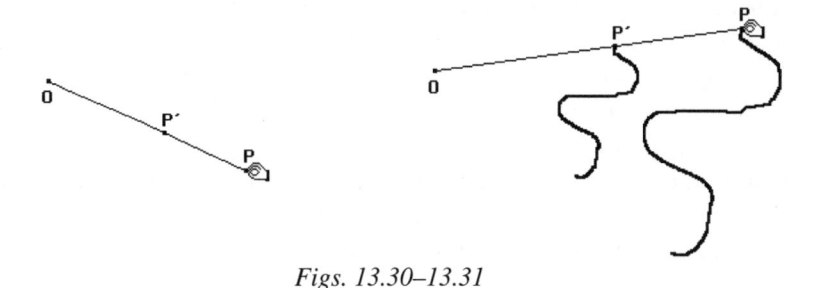

Figs. 13.30–13.31

13.7 GLIDE REFLECTION

A microworld for glide reflection is not discussed here and is left as an exercise for the reader. (Macros for TRANSLATION, ROTATION, LINE REFLECTION, GLIDE REFLECTION and ENLARGEMENT are supplied on the disc associated with this book.)

CHAPTER 14

Theorem finding – an advanced case study

Eureka!
Archimedes of Syracuse

14.1 INTRODUCTION

Which fundamental insights should be gained by older students in theorem finding or theorem discovery in geometry? We suggest the following.

The students should:

- learn that geometrical activities may lead to the rediscovery (i.e. the personal discovery) of already known theorems;
- understand that there are many yet undiscovered theorems in geometry that might, under certain circumstances, be discovered by the students (although the 'less complex' theorems are very likely to have been discovered already);
- realise that measurements and constructive tests of geometrical propositions on drawings on paper or screen cannot provide a *definite proof* of the general validity of these propositions.

Additionally the following skills should be achieved by the process of theorem finding.
The students should be able to:

- use preliminary knowledge of geometrical facts in order to arrive at a conjecture;
- interpret geometrical data from constructions and measurements in order to find and confirm or refute a conjecture;
- recognise precisely which geometrical information is needed to establish a conjecture;

- distinguish between interesting and uninteresting conjectures (Is the conjecture already known to be a theorem? Can the assumed proposition be generalised or specialised? Can the conjecture be immediately derived from already known theorems?...);
- arrange conjectures (obtained by exploration) in an order that corresponds to their assumed logical relationship;
- formulate a conjecture involving shapes, properties and relations using such terms as "if ... then ...", or "all ... are ...", etc.

Communicative and affective objectives will not be mentioned here, which is not to deny their importance.

A classification of theorem finding or theorem discovery as part of a global treatment of geometrical theorems is to be found in Chapter 4 *Discovering theorems by varying geometrical figures*, where many examples are given explaining how theorem finding processes can be supported by interactive graphic software.

A process of interactive theorem finding
As a general rule initial conjectures about screen drawings are obtained by visual evidence on the basis of one's current knowledge of geometry. These conjectures refer to relations between objects or to properties of objects, which are part of the drawing on the screen. Continuous variation of the drawing allows one to test whether properties are invariant (confirmation of the conjecture) or not (rejection of the conjecture, by counter-example). Initial conjectures can additionally be confirmed or rejected by interactive measurements or simulation of ruler-and-compasses constructions and generation of loci). This process is not generally linear but takes place in a spiral. It should conclude with documentation supporting the empirically verified conjecture, in the form of text and drawing, possibly using screen dumps.

In Section 14.2 below a real example of interactive theorem finding is reported, which was supported – or rather enabled – by Cabri-géomètre.
(Note: The **Edit** tools used included: **Look of objects** in order to conceal auxiliary lines and to highlight objects, **Name** for labelling objects; the **Construction** tools included: the system macros **Midpoint, Perpendicular line, Parallel line, Centre of a circle**, and various user-defined macros such as 3-POINT-CIRCLE.)

14.2 A MODEL OF THEOREM FINDING – A CASE STUDY

We construct a cyclic quadrilateral ABCD inscribed in a circle with centre M. From each side we raise a perpendicular which passes through the midpoint of the opposite side. These four perpendiculars intersect in a single point S (Fig. 14.1). When we change the position on the circle of A or B or C or D the property is preserved. This property is well-known; it is even true for a self-intersecting cyclic quadrilateral (Fig. 14.2). As a vertex is moved, S moves and can be made to coincide with the midpoint of either of those two sides which have no point in common with the dragged vertex (Fig. 14.3, where vertex A is varied; S can lie on the midpoint of BC or the midpoint of AD.) This concurrence of S and a midpoint occurs when the midpoint of the opposite side coincides with M, the centre of the circumcircle of ABCD. In Fig. 14.3, S coincides with the midpoint of CD, and the midpoint of AB coincides with M.

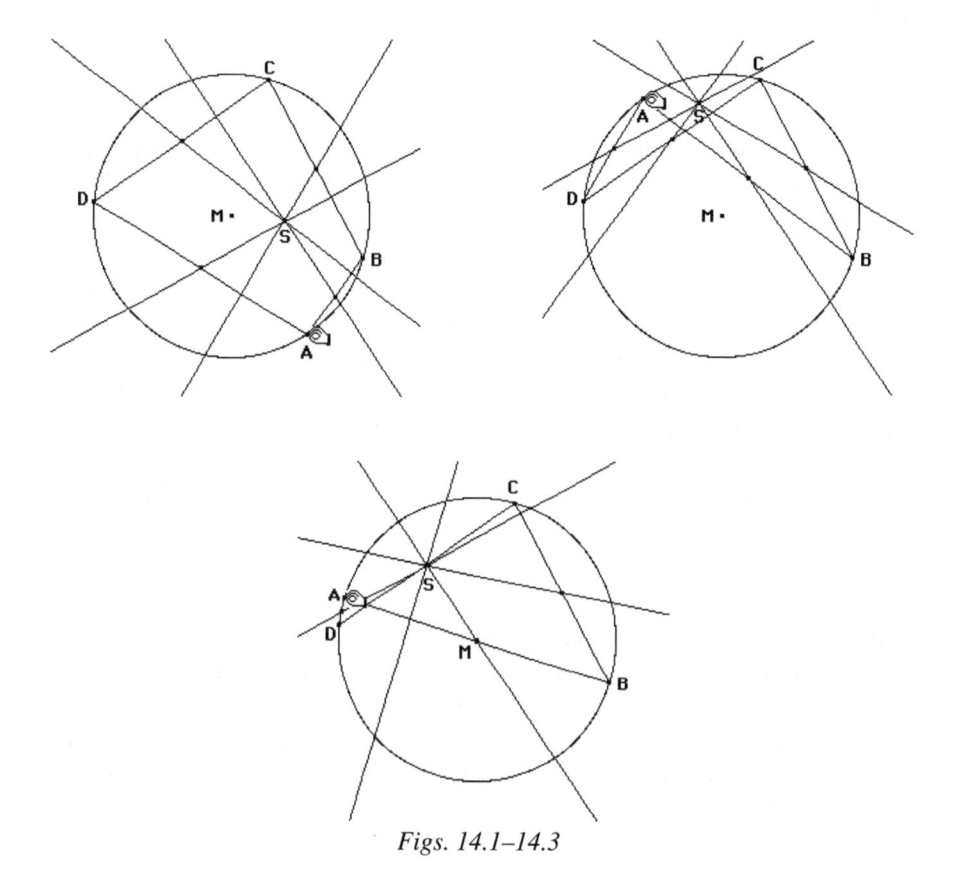

Figs. 14.1–14.3

Theorem finding – an advanced case stydy

Question: "What is the locus of S as one vertex travels right round the circumcircle?"

Interactive generation of the locus of S produces a circle (Fig. 14.4), uniquely determined by S and the midpoints of the two sides opposite the dragged vertex. We construct, with the help of the macro 3-POINT-CIRCLE, a circle from these two midpoints and S (Fig. 14.5), which coincides with the generated locus. The size of this circle is not changed by variation of the quadrilateral ABCD – this can be verified empirically by measurement (Fig. 14.6). The same result is then obtained for the other three circles through the midpoints of neighbouring sides and S (Fig. 14.7).

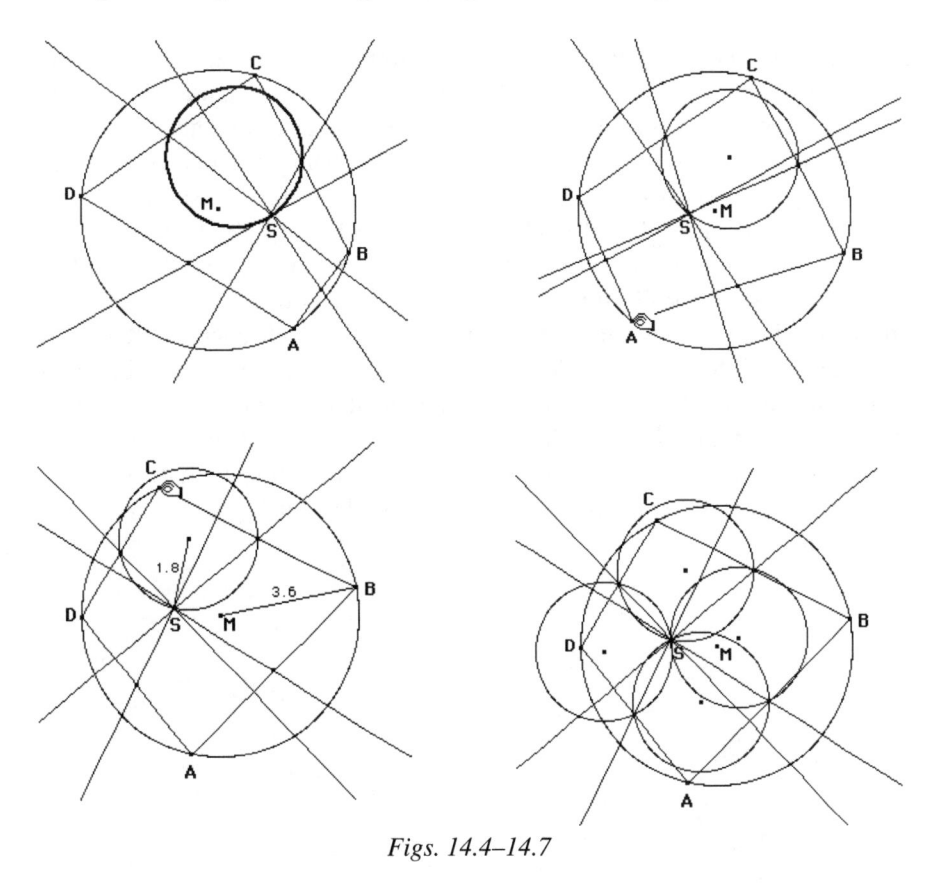

Figs. 14.4–14.7

Main Theorem: The centres of the four circles form a quadrilateral, which is similar to the cyclic quadrilateral ABCD, and its sides are parallel to the sides of ABCD (Fig. 14.8).

253

Fig. 14.8

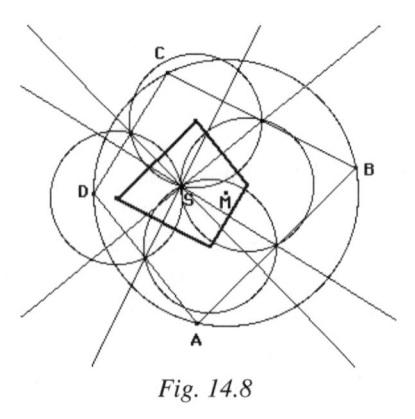

This can be verified empirically by measurement or construction. In Fig. 14.9 only some measurements are shown and just one parallel-checking construction is drawn in. The sides of the quadrilateral have half the length of the sides of ABCD, the corresponding sides being parallel. ABCD and A'B'C'D' are homothetic i.e. found in similarly concentric positions, so one is an enlargement of the other about a point. In Fig. 14.10, the scale factor is −1/2, the centre of enlargement is Z which lies on the line through SM such that |SM| = 3 |SZ|. In Fig. 14.11 Z is hidden. A'B'C'D' is a cyclic quadrilateral, because it is similar to the initial cyclic quadrilateral ABCD. S is the centre of the circumcircle of A'B'C'D', since the perpendicular lines through S are at the same time the perpendicular bisectors of the sides of A'B'C'D'. Varying the quadrilateral ABCD does not change these results (Fig. 14.12). Clearly we can repeat the process starting with A'B'C'D' to produce A"B"C"D" (Fig. 14.13), and so on, perhaps using a macro defined in advance.

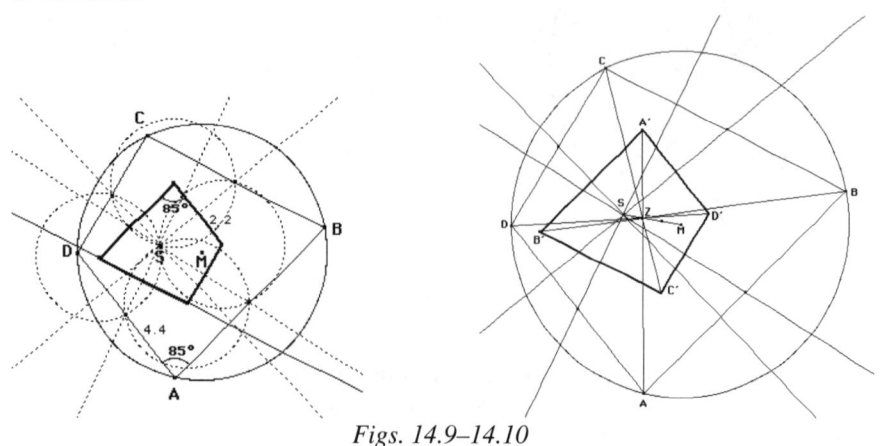

Figs. 14.9–14.10

Theorem finding – an advanced case stydy

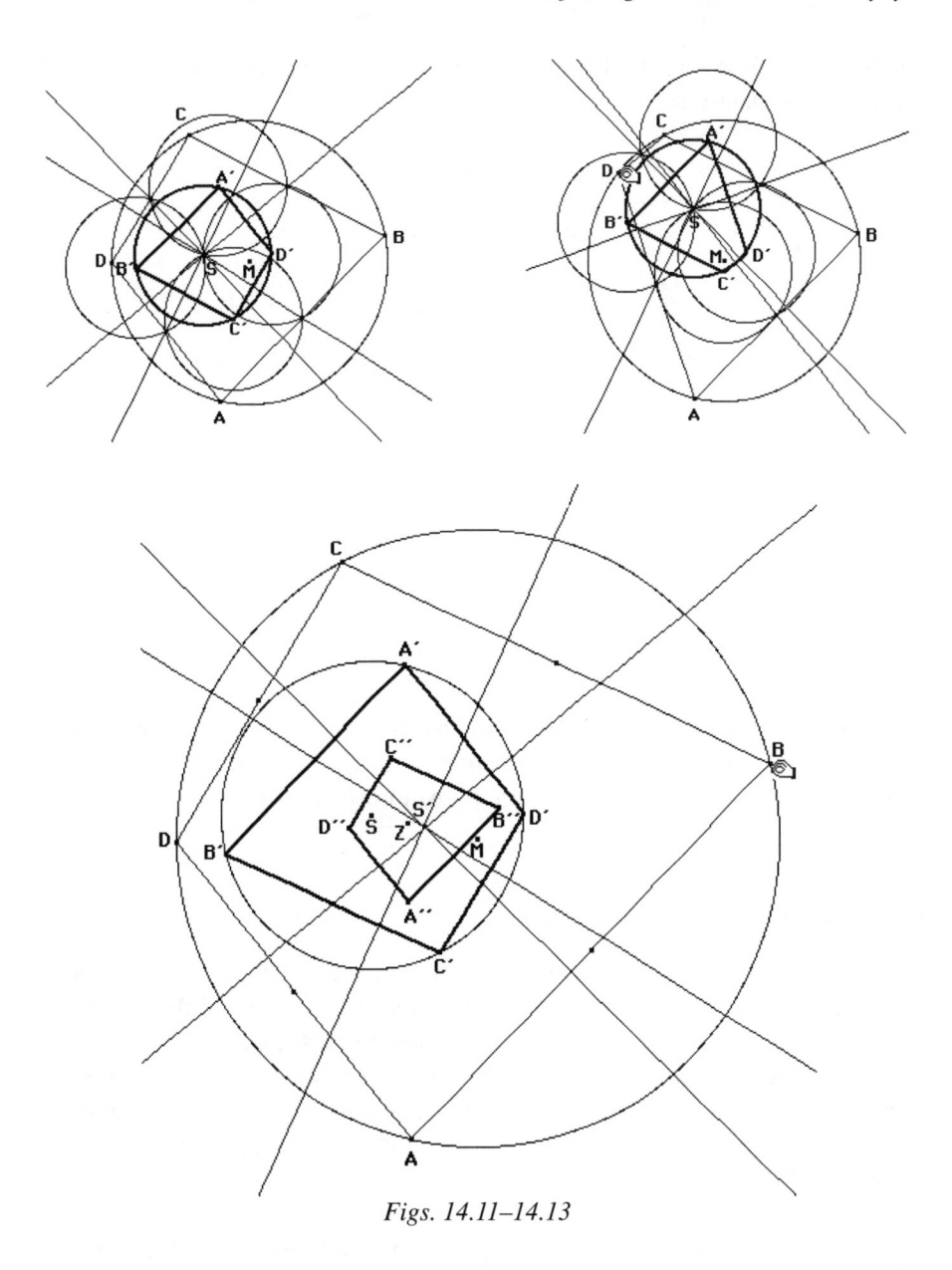

Figs. 14.11–14.13

The reader can confirm that S' satisfies |S'S| = |S'M|. Z remains the centre of the homothetic transformation; the points of intersection S, S', S" ... of corresponding perpendiculars lie on a straight line; the sequence of points of intersection of perpendiculars converges 'alternatingly' towards Z; the distance between the points of intersection of the perpendiculars and Z is halved each time. Fig. 14.14 shows a converging sequence of nested similar cyclic quadrilaterals.

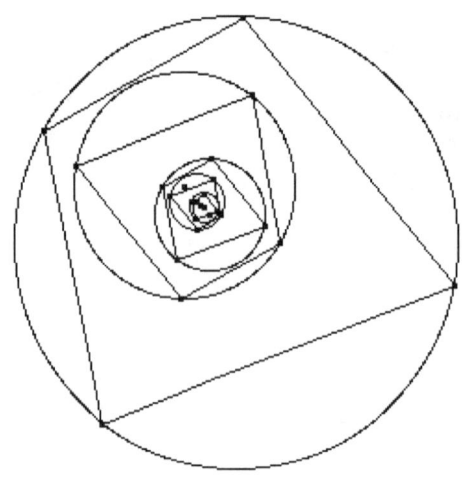

Figs. 14.14

The main theorem and the detailed results, which were recently rediscovered by one of the authors using Cabri-géomètre, were reported in essence in the *Mathematical Gazette* by Ambrose (1966).

14.3 ADDITIONAL REMARKS

This example proves the efficiency of interactive tools like Cabri-géomètre for the discovery of new facts in plane geometry. The original discovery of such facts may be reproduced with traditional construction tools, but certainly is not likely to be initiated in that way. A proof of the main theorem in Section 14.2 has been accomplished using vector algebra and similarity transformations but a synthetic proof using elementary geometry methods remains elusive.

Apart from the possibilities of empirical methods of proof, the current Macintosh version of Cabri-géomètre offers a special semi-intelligent facility for direct (empirical) testing of hypotheses. With the command **Check property** (Fig. 14.15) the ternary relation **Alignment** (collinearity of three points) and the binary relations **Membership** (point on line or circle), **Parallelism** (two lines), **Perpendicular lines** (two lines) and **Equal**

Theorem finding – an advanced case stydy

lengths (two lines) can be checked by clicking to indicate the objects to be tested. With the current versions of Cabri-géomètre, commands for checking equality of angles, areas etc. are not available.

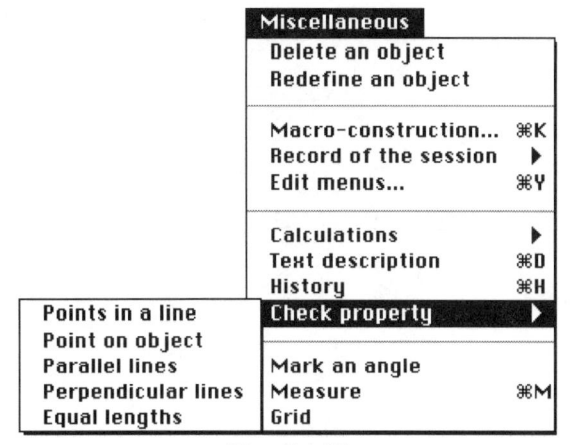

Figs. 14.15

Cabri-géomètre checks internally whether the numerical-analytical assumptions for the relational properties of the screen drawing are true. It does this by slightly varying the figure to see if the property in question is invariant; the precise nature of the algorithm is not explicable in simple terms. This test depends on the internal accuracy of Cabri-géomètre and it appears that the probability of rejecting a valid theorem can be reduced to any desired level with increasing arithmetic accuracy. If variation is found then Cabri-géomètre can present a counter-example. There are very small probabilities of rejecting a true theorem and of accepting a false theorem but, unlike normal statistical processes, theorem generation is not stochastic in nature and so it is not known whether such errors will occur in fact. If Cabri-géomètre tells us that a property is true (Fig. 14.16) then this means: "there is a high probability that this property is true".

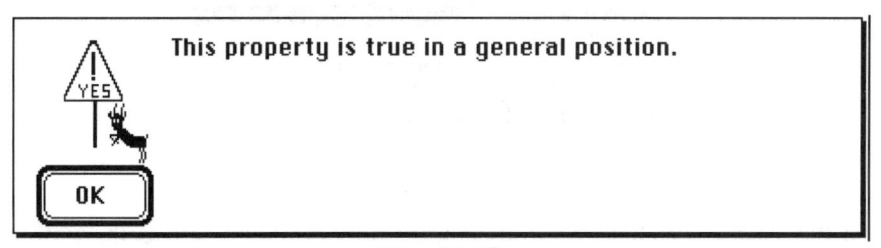

Figs. 14.16

Discovering Geometry with a Computer

With reference to Fig. 14.17 (a variant of Fig. 14.6), it is possible to check, for instance, whether the radius of the circle through the midpoints of the two neighbouring sides and S is half the radius of the circumcircle of the cyclic quadrilateral ABCD.

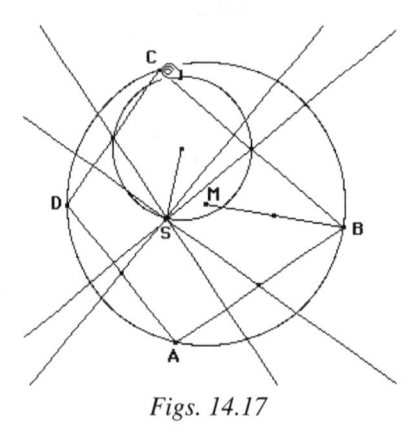

Figs. 14.17

After selecting **Equal lengths** (Fig. 14.18) the radius of the smaller circle and half the radius of the circumcircle (MB) are selected by appropriate clicks on the screen drawing. (Note that only *equal* lengths can be checked so it is necessary here to bisect MB first in order to test the equality of half of MB and the smaller radius.)

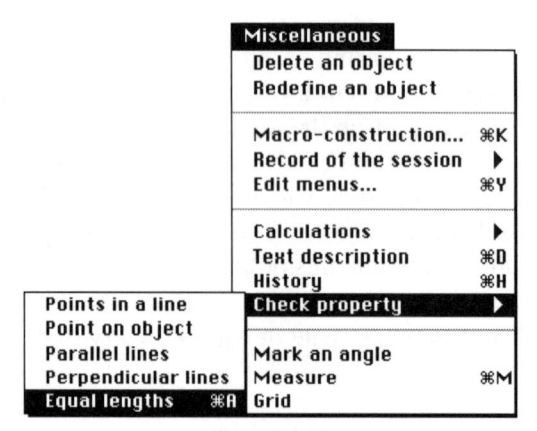

Figs. 14.18

In the case of an incorrect conjecture, Cabri-géomètre rejects the untrue property and offers a counter-example. The basic objects of the screen drawing are moved in a way that should make the user recognise that the conjecture is untrue. Obviously, the availability of this property checking facility is likely to reduce a student's motivation to find a rigorous proof argument to confirm a conjecture. Furthermore, the authors have up to now been unable to find an example where the property checking command of Cabri-géomètre fails. Therefore we have to face the problem of presenting 'proving' as a fundamental mathematical skill in geometry in a way which is suited to motivate the students. One way is by emphasising the necessity to classify the great number of interactively discovered theorems with the purpose of setting up a hierarchy of theorems. In the upper grades of school and at college a degree of strictness in proof arguments may be appropriate – based upon deductions from established theorems. But what happens if an Euclidean proof – as in our example – cannot be found? The balance between inductive and deductive cognition is disturbed. From the epistemological point of view, we are facing a risk of a throwback to pre-Euclidean times when there was mainly inductive establishing of geometrical statements and their application without formal proof. It will be a loss if able students simply use tools like Cabri-géomètre and do not think about proofs at all. However, proofs figure very little in today's curricula anyway so Cabri-géomètre cannot be seen as a threat. Indeed it may provide a stimulus to a revival in formal proof.

CHAPTER 15

A geometric story

A picture is worth a thousand words.
Confucius

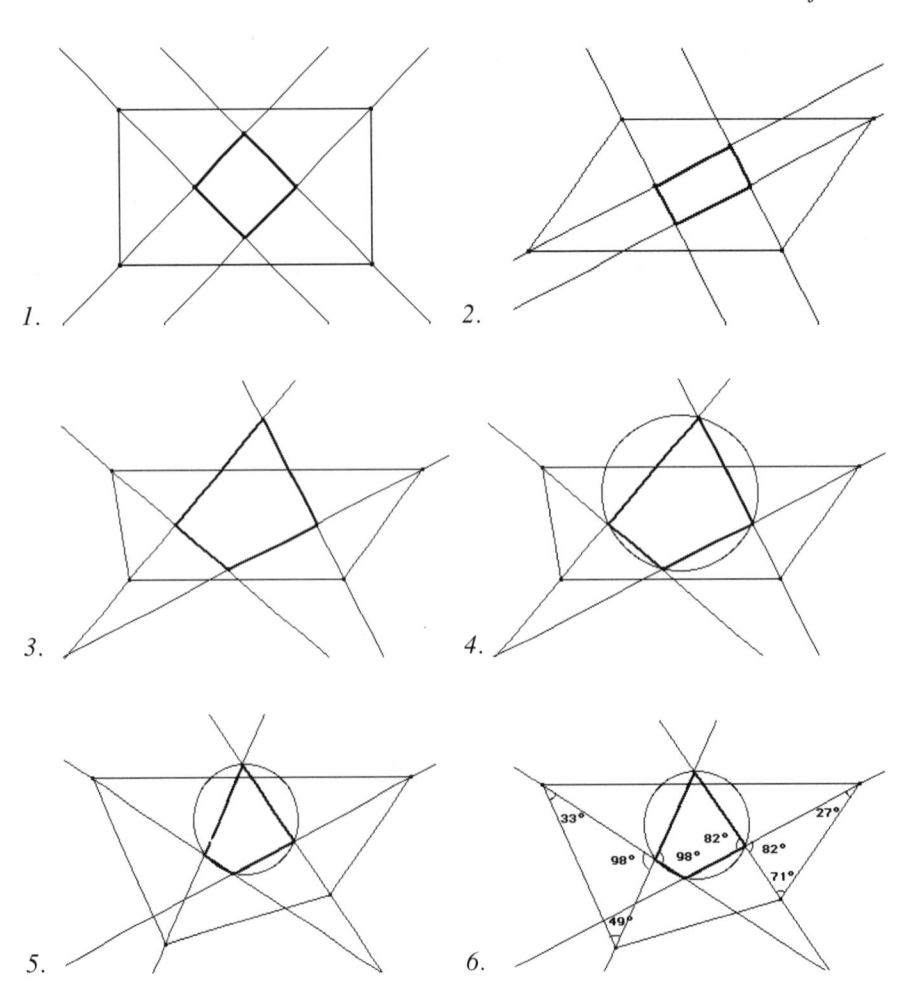

260

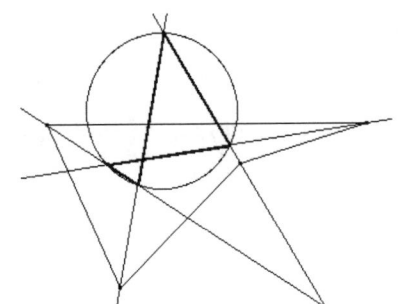

A geometric story

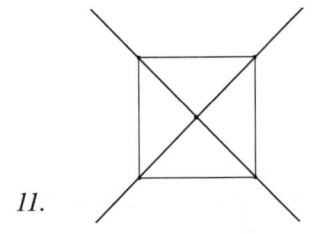

7.

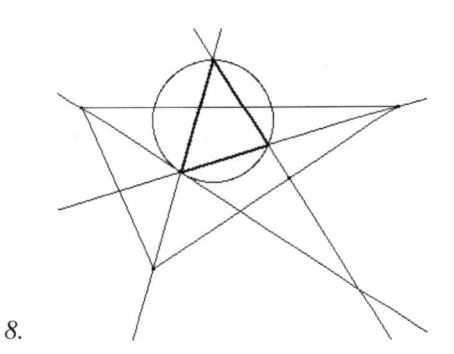

8.

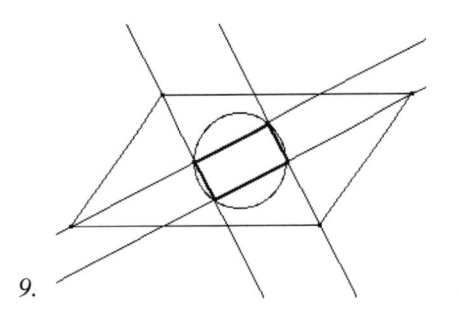

9.

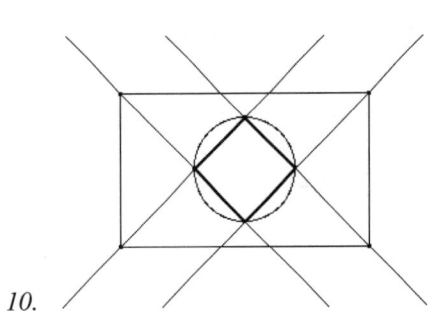

10.

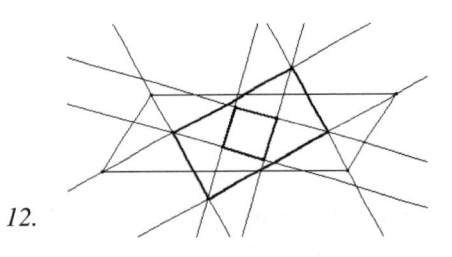

11.

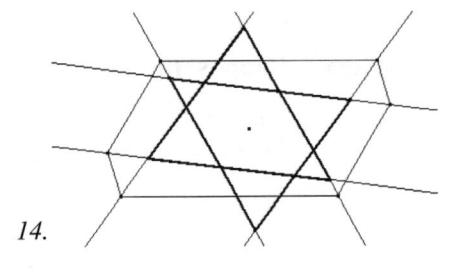

12.

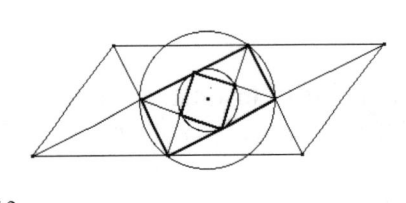

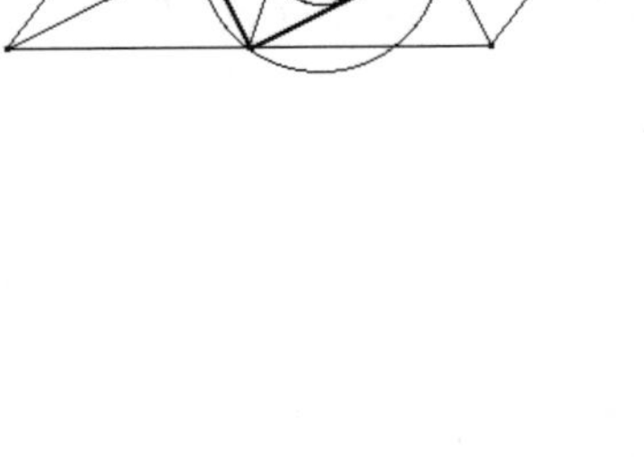

13.

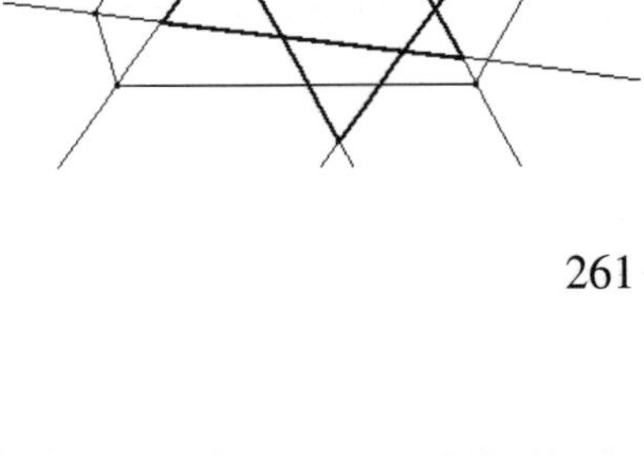

14.

Appendix

Twenty problems to investigate

Lines

Problem 1 (Subdividing a line segment)

Construct a line segment AB and a line through A at an arbitrary angle to AB (say 30°). Place a point P_1 on the line through A and by drawing a circle centre P_1 radius AP_1 find a point P_2 on the line such that $AP_2 = 2AP_1$. Repeat the process to find P_3 such that $AP_3 = 3AP_1$.

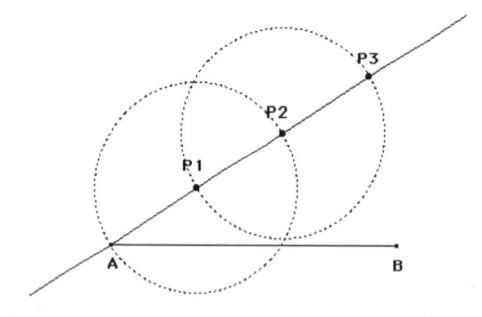

Draw the line segment BP_3 and draw two lines parallel to BP_3 passing through P_2 and P_1. Investigate the positions of the points of intersection of those two lines and AB. Generalise ... (Note: if making a macro for this construction, to avoid having an *arbitrary* position for P_1 one could take AP_1 equal in length to AB.)

Problem 2 (Pappus' theorem)

Points are joined between two lines as indicated. Investigate the property possessed by the three intersection points: AC' and A'C, AB' and A'B, BC' and B'C.

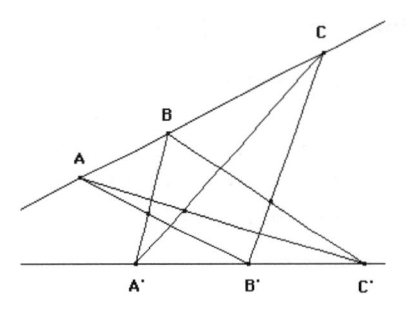

Triangles

Problem 3 (Pythagoras' inverse square law?)

A right-angled triangle ABC with hypotenuse AB has squares erected outwards on the two shorter sides and inwards on the hypotenuse (i.e. crossing over the triangle). What do you notice?

Problem 4 (Trisection triangle area)

In a triangle ABC lines are joined from the vertices to points trisecting the opposite sides as in the diagram.

(a) Investigate the area of the inner triangle in relation to the area of triangle ABC.
(b) Investigate inner triangles made using other proportional positions.

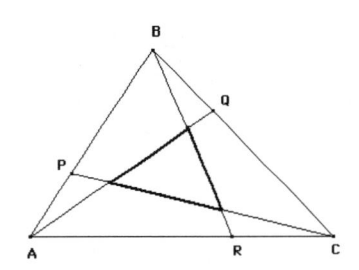

Problem 5 (Midpoint symmetric triangle areas)

A triangle is drawn and the midpoints of its sides found. Pairs of points are placed on each side of the triangle (P_i, Q_i), symmetric about the midpoint of the side (i.e. three free points and three dependent points). Corresponding points are joined to form two triangles ($P_1P_2P_3$ and $Q_1Q_2Q_3$) as in the diagram. Investigate the relationship between their areas.

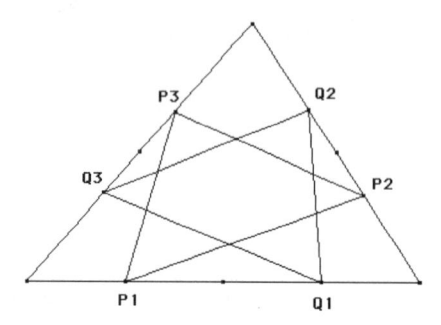

Problem 6 (Medians)

Two medians BB' and CC' of triangle ABC are drawn and meet at G. The midpoints L and M of GB and GC are found. What shape is LMB'C'? (Find a proof.) Use this result to prove that the medians trisect each other.

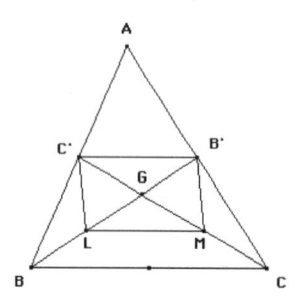

Problem 7 (The Euler line)

Verify that the circumcentre, centroid and orthocentre of a triangle are collinear (lying on the Euler line).

Problem 8 (Based on a question form the 1975 Mathematical Olympiad)

ABC is an arbitrary triangle and points PQR are such as to give the angles shown in the figure below. Find the relationship between QR and RP and determine the angle QRP.

(One elegant approach involves finding two triangles in the figure such that one is the image of the other under rotation, which immediately yields both answers. Glaeser (1986) reports that this method of proof was not that intended by the question setter but one competitor discovered it much to everyone's surprise.)

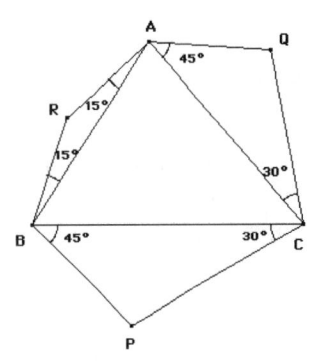

Circles

Problem 9 (Johnson's theorem)

Three identical circles are drawn to pass through a common point P. What can you discover about the other points of intersection?

Problem 10 (Three touching circles and a common tangent)

Three circles with a common tangent all touch each other. Devise a method of drawing such a figure. Investigate its properties.

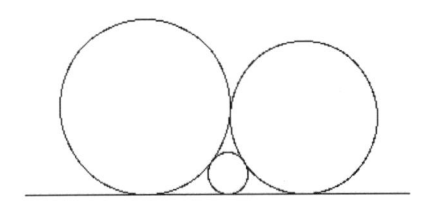

Problem 11 (Miquel's theorem)

Draw a circle and mark on it four points A, B, C, D. Draw four other circles of arbitrary sizes, one through A and B, one through B and C, one through C and D, and one through D and A. What can you discover about the four other points of intersection of the four circles?

Problem 12 (The arbelos – a theorem of Pappus)

Inside a large semi-circle, two smaller semi-circles are drawn touching each other and touching the large semi-circle, all three diameters being on a common line. The region bounded by the three semi-circles was named by Archimedes the *arbelos* – meaning shoemaker's knife – due to its shape.

A sequence of small circles $c_1, c_2, c_3, \ldots$ is inscribed in the arbelos, as illustrated in the diagram.

(a) Investigate the relationship between the diameter of each small circle, c_i, and the height of its centre above the diameter of the large semi-circle. (A table with columns headed i, c_i, height$_i$ may help.)

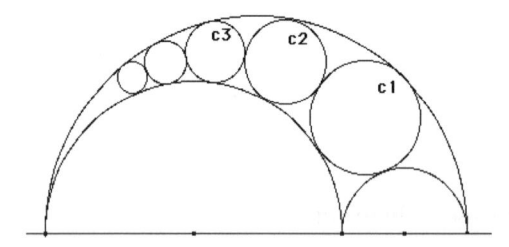

(b) What is the situation if the right semi-circle is omitted and circle c_1 touches the main diameter?

See Wells (1991) and Cadwell (1966) for many fascinating details of the arbelos problem.

Triangles and Circles

Problem 13 (The Nine-point circle)

Verify the following properties:

(a) In any triangle the midpoints of the sides, the feet of the altitudes and the midpoints of the lines joining the orthocentre to the vertices all lie on a circle.
(b) The centre of the Nine-point circle is midway between the orthocentre and the circumcentre.
(c) The radius of the Nine-point circle is half the radius of the circumcircle.
(e) An enlargement with centre the orthocentre and scale factor 2, takes the Nine-point circle onto the circumcircle.
(e) The Nine-point centre lies on the Euler line (see Problem 7).

Problem 14 (Wallace's theorem and the Miquel point)

Any four lines will in general intersect to make four triangles, as in the diagram.

What property do the four circumcircles of such a set of triangles have?

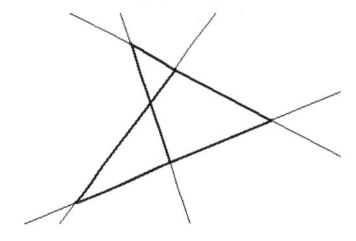

Problem 15 (The Simson line)

The circumcircle of a triangle ABC is drawn. A free point P is placed on the circumcircle and perpendiculars are dropped from P to the sides of the triangle (extended if necessary) meeting the sides at P_a, P_b, P_c. What relationship do P_a, P_b, P_c have? Is P_a always between P_b and P_c?

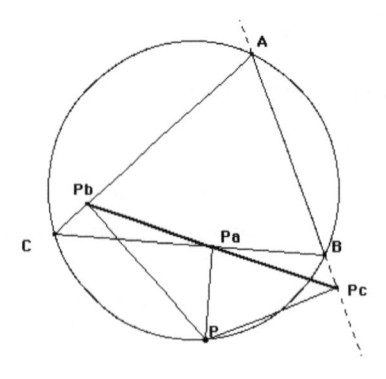

Polygons

Problem 16 (Inscribing square problem)

Using the figure below for inspiration devise a method for inscribing a square in a triangle. When is the method valid and when not?

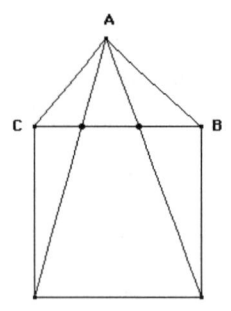

Problem 17 (Octagon on trial)

The midpoints and vertices of a square are joined as indicated to make a series of triangles which enclose an octagon. Is it equilateral? Is it regular?

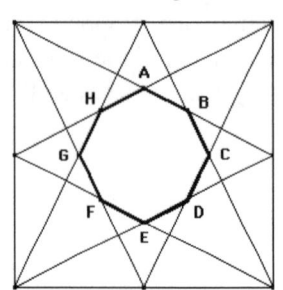

Loci

Problem 18 (Loci of a foot on the Simson line)

The Simson line was the subject of Problem 15. P_a, P_b, P_c are three points on the Simson line, being the feet of the perpendiculars from a point P on the circumcircle to the sides of the triangle.

(a) Find the locus of P_a as P moves round the circumcircle.
(b) Find the locus of P_a as B moves round the circumcircle.
(c) Find the locus of P_a as C moves round the circumcircle.
(d) Find the locus of P_a as A moves round the circumcircle.

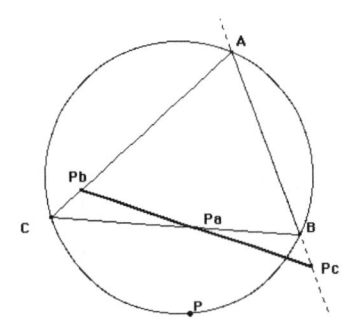

Problem 19 (Loci of midpoint of two feet on the Simson line)

M_{bc} is the midpoint of the feet P_b, P_c on the Simson line.

(a) Find the locus of M_{bc} as vertex A moves round the circumcircle.
(b) Find the locus of M_{bc} as vertex B moves round the circumcircle.
(c) Find the locus of M_{bc} as vertex C moves round the circumcircle.
(d) Find the locus of M_{bc} as point P moves round the circumcircle.

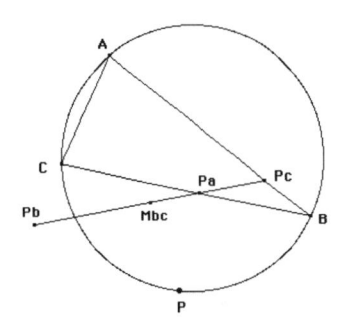

Problem 20 (Based on a idea by Adrian Oldknow, M.A. Conference, Easter 1994)

On a line are placed free points A and O and a circle centre O radius OA is drawn. A free point P is placed on the line, initially between A and O. A second free point is placed on the circle and PQ is drawn. A chord to the circle is drawn from Q perpendicular to PQ and its midpoint, M, is found. Investigate the locus of M as Q travels round the circle, for various positions of P.

Suggested positions for P are:

(a) midway between A and O.
(b) at O
(c) 'near' O
(d) at A
(e) 'near' A
(f) outside but 'near' the circle
(g) outside and 'far' from the circle.

(Try to come up with a comprehensive theory by considering the question "How does the locus change as P moves from O to A?")

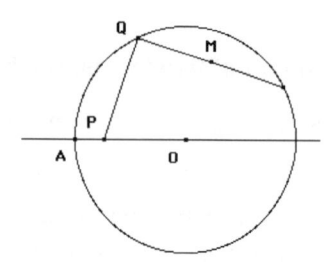

Notes on Problems

1. This gives a standard construction for subdividing a line segment into equal sub-segments.
2. Pappus' theorem states that the three intersection points are collinear.
3. A vertex of the large square always lies on a side of a smaller square (extended if necessary).
4. The area of the inner triangle is 1/7 that of the outer triangle.
5. The areas of the two triangles are equal.
6. By similar triangles applied twice LMB'C' is a parallelogram. The property that the diagonals of a parallelogram bisect each other gives the required result.
7. A standard result found in textbooks.
8. RP and QR are equal in length and angle QRP is 90°.
9. They lie on a circle the same size as the other circles. (Discovered by Johnson in 1916.)
10. No help provided!
11. They lie on a circle.
12. (a) $\text{height}_i = i \times c_i$ (b) $\text{height}_i = (2i - 1) \times c_i$
13. A standard result found in textbooks.
14. Wallace's theorem states that the four circumcircles are concurrent. They meet at the Miquel point.
15. P_a, P_b, P_c are collinear – lying on the Simson line. Their order depends upon the position of P and the positions of the vertices of the triangle.
16. Ideas of perspective or enlargement may help.
17. It is equilateral but not regular. The irregularity is seen by considering the lengths of the sides and base of the triangle with sides passing through HG and BC, and comparing with those of the triangle with sides passing through AH and CD.
18. (a) line segment overlapping BC. (b) circle through P, P_a, C
    (c) circle through P, P_a, B (d) a fixed point (A and P_a are independent).
19. (a) circle through P, P_a (b) circle through P_b
    (c) circle through P_c (d) ellipse through A.
20. No help provided!

GLOSSARY

(for the non-expert in geometry)

Absolute geometry	– a system based on Euclid's first four postulates *without* the fifth postulate (which asserts the unique parallel to a line through a point)
Affine geometry	– a system based on Euclid's first two and the fifth postulates *without* the third and fourth postulates (which concern circles and angles)
Altitude	– perpendicular line segment from a vertex of a triangle to the opposite side (extended if necessary)
Bisector	– that which cuts into two equal parts
Cardioid	– a limaçon with polar equation $r = 2a \cos\theta + 2a$
Centroid (G)	– the common point of intersection of the medians of a triangle
Chord	– a line segment joining two points on the circumference of a circle (dividing the circle into two parts)
Circumcentre (O)	– centre of circumcircle
Circumcircle	– circle passing through the three vertices of a triangle
Cissoid	– curve with polar equation $r^2 = 2a \sin^2\theta / \cos\theta$
Collinear	– lying on a common line
Complementary angles	– a pair of angles which sum to 90°
Conchoid (of Nicomedes)	– curve with polar equation $r = a \sec\theta + k$

Glossary

Conic	– the locus of a point whose distance from a fixed point (the focus) is a multiple e (the eccentricity) of its distance from a fixed line (the directrix): $0 < e < 1$ for ellipse; $e = 1$ for parabola, $e > 1$ for hyperbola.
Conic section	– a conic (so named because different plane sections through a cone give these particular curves)
Congruent	– identically equal in both size and shape (i.e. congruent figures have corresponding angles equal and corresponding sides equal)
Cyclic quadrilateral	– quadrilateral whose four vertices all lie on a circle
Directrix	– the fixed line of a conic
Eccentricity	– the value of e for a conic
Ellipse	– a conic for which $0 < e < 1$
Equilateral	– having all sides of equal length
Escribe	– draw a figure outside another figure such that the vertices of one are in contact with the side(s) of the other
Excentre (I_a, I_b, I_c)	– centre of an excircle (a triangle has three)
Excircle	– circle touching one side of a triangle and the two other sides extended (i.e. outside the triangle)
Focus	– the fixed point of a conic
Foot	– where an altitude meets the side of a triangle
Hyperbola	– a conic for which $e > 1$
Hypotenuse	– the side of a right-angled triangle opposite the right-angle
Incircle	– circle touching the three sides of a triangle (i.e. inside the triangle)
Incentre (I)	– centre of incircle
Inscribe	– draw a figure inside another figure such that the vertices of one are in contact with the side(s) of the other
Inversion	– a transformation which preserves angles but transforms (some) lines into circles and vice versa. Given a fixed circle, with centre O, and radius k, the inverse of any point P

	(except the point O itself) is the point P' such that $OP.OP' = k^2$
Limaçon	– curve with polar equation $r = 2a \cos\theta + k$
Lemniscate	– curve with polar equation $r = b + a \sin\theta$ *or* $r = b + a \cos\theta$
Locus	– the path of a moving point *or* the set of all possible positions which a point satisfying some condition(s) can occupy
Median	– line segment from a vertex of a triangle to the midpoint of the opposite side
Midpoint	– point bisecting a line segment, or midway between two points
Normal	– line perpendicular to the tangent at a point on a curve
Orthic triangle	– triangle formed by the three feet of the altitudes of a triangle
Orthocentre (H)	– the common point of intersection of the altitudes of a triangle
Orthogonal	– at right-angles
Parabola	– a conic for which $e = 1$
Perpendicular	– at right-angles (also a line at right-angles to another)
Pedal triangle	– orthic triangle
Regular	– having all sides and angles of equal magnitude
Similar	– similar figures have corresponding angles equal and corresponding sides in proportion
Supplementary angles	– a pair of angles which sum to 180°
Subtend	– to be opposite to, to extend under
Trapezium	– a quadrilateral with precisely one pair of parallel sides
Trisectrix	– a limaçon with polar equation $r = 2a \cos\theta + a$
Trisectrix of Maclaurin	– curve with polar equation $r = a \sec(\theta/3)$
Vertex	– a point where two sides of a polygon meet (i.e. a 'corner') *or* a point where a conic intersects its major axis

BIBLIOGRAPHY

Ambrose, D.P. (1966) Three "Eight–Point" circles of a cyclic quadrilateral. *Mathematical Gazette,* **50**(373), 301–3.

Anderson, J.A. et al. (1986) The Geometry Tutor. *The Journal of Mathematical Behavior,* **5**, 5–19.

Bolt, A.B. and Hiscocks, J.E. (1970) *Machines, Mechanisms and Mathematics.* London: Chatto and Windus.

Cadwell, J.H. (1966) *Topics in Recreational Mathematics.* Cambridge: C.U.P.

Chazan, D.and Houde, R. (1989) *How to Use Conjecturing and Microcomputers to Teach Geometry.* Reston, VA: National Council of Teachers of Mathematics.

Chou, S-C. (1988) *Mechanical Geometry Theorem Proving.* Dordrecht: D Reidel Pub. Co.

Court, N.A. (1952) *College Geometry 2nd Edition.* New York: Barnes and Noble.

Coxeter, H.S.M. (1961) *Introduction to Geometry.* New York: Wiley.

Cundy, H.M and Rollett, A.P. (1961) *Mathematical Models, 2nd Edition.* Oxford: O.U.P.

Fisher, J.C. et al. (1981) Polygons and Polynomials, *in* Davis, C. et al. *The Geometric Vein.* New York: Springer-Verlag.

Glaeser, G. (1986) The crisis of geometry teaching, *in* Morris, R. (ed.), 107-122.

Grünbaum, B and Shephard, G.C. (1989) *Tilings and Patterns – an introduction.* New York: W H Freeman.

Heath, T.L. (1956) *The Thirteen Books of Euclid's Elements, 3 Vols.* New York: Dover.

Henderson, K. ed. (1973) *Geometry in the Mathematics Curriculum: 1973 Yearbook.* Reston, VA: National Council of Teachers of Mathematics.

Hilbert, D. (1902) *Foundations of Geometry.* Chicago: Open Court.

Hoffer, A. (1978) Geometry is more than proof. *Mathematics Teacher,* **74,** 11–18.

Hudson, H.P. (1953) *Ruler and Compasses.* New York: Chelsea.

Iaglom, I.M. (1962, 1968) *Geometric Transformations, 2 Vols.* Washington DC: Mathematical Association of America.

Jeger, M. (1966) *Transformation Geometry.* London: Allen and Unwin.

Johnson, R.A. (1960) *Advanced Euclidean Geometry.* New York: Dover.

Klein, F. (1939) *Elementary Mathematics from an Advanced Standpoint, Vol. 2.* New York: Macmillan.

Kostovskii, A.N. (1961) *Geometrical constructions using compasses only.* Oxford: Pergamon Press.

Laborde, J-M. (1988) A propos de la validation de faits géométriques dans un systéme d'enseignement assisté par ordinateur dans la domaine de la géométrie élémentaire – *Rapport Recherche.* LSD (IMAG), Université Joseph Fourier, Grenoble.

Lindquist, M. ed. (1987) *Learning and teaching geometry, K–12: 1987 Yearbook.* Reston, VA: National Council of Teachers of Mathematics.

Lockwood, E.H. (1967) *A Book of Curves.* Cambridge: CUP.

Maxwell, E.A. (1975) *Geometry by Transformations.* Cambridge: CUP.

Moore, O.K., Anderson, A.R. (1969) Some Principles for the Design of Clarifying Educational Environments, *in* Goslin, A. (ed.) *Handbook of Socialization Theory and Research.* Chicago: Rand McNally.

Morris, R. (1986) *Studies in Mathematics Education Vol 5: Teaching of Geometry.* Paris: UNESCO.

Norman, D.A. (1983) Some observations on mental models, *in* Gentner, D., Stevens, A.L. (Eds.) *Mental Models.* Hillsdale, NJ: L. Erlbaum Associates.

Ogilvy, C.S. (1969) *Excursions in Geometry.* New York: OUP.

Pedoe, D. (1957) *Circles – A Mathematical View.* New York: Dover.

Pedoe, D. (1970) *A Course of Geometry.* Cambridge: CUP.

Perfect, H. (1963) *Topics in Geometry.* Oxford: Pergamon.

Polya, G. (1957) *How to Solve It.* Garden City, New York: Doubleday.

Robitaille, D.F. and Travers, K.J. (1986) Geometry for 13-year-olds in Canada and the United States, *in* Morris, R. (ed.), 23–30.

Schumann, H. (1988) Der Computer als Werkzeug zum Konstruieren im Geometrieunterricht. *ZDM (International Reviews on Mathematical Education),* **20**(6), 248–263.

Schumann, H. (1989) The computer as a tool for geometric constructions. *Micromath,* **5**(3), 53–6.

Schumann, H. (1991) Interactive Theorem Finding through Continuous Variation of Geometric Configurations, *Journal of Computers in Mathematics and Science Teaching.* **10**(3), 81-105.

Schumann, H. (1991) *Schulgeometrisches Konstruieren mit dem Computer.* Stuttgart: Teubner/Metzler.

Schumann, H. and Strässer, R. (eds.) (1992) Analysen: Computerunterstützer Geometrieunterricht. *ZDM (International Reviews on Mathematical Education),* **24**(5/6), 119-153/165–203.

Schumann, H. and de Villiers, M. (1993) Continuous variation of geometric figures: interactive theorem finding and problems in proving. *PYTHAGORAS,* **31** (April), 9–20.

Schumann, H. (1995) Interactive calculations on geometric figures. *International Journal of Mathematical Education in Science and Technology,* **26**(2).

Schwartz, J.L., Yerushalmy, M. (1985) *The Geometric Supposer: Triangles.* Pleasantville, NY: Sunburst Communications.

Taylor, R. (1980) *The Computer in the School.* New York: London.

Wells, D. (1991) *The Penguin Dictionary of Curious and Interesting Geometry.* London: Penguin Books.

Weyl, H. (1952) *Symmetry.* Princeton: Princeton University Press.

Willson, W.W. (1977) *The Mathematical Curriculum: Geometry.* Glasgow: Blackie.

Yerushalmy, M. (1986) *Induction and Generalization: An Experiment in Teaching and Learning High School Geometry.* (thesis) Ann Arbor: UMI.

Index

Note Cabri-géomètre commands and menus are shown **bold** and macro names are capitalised.

accuracy 90
affine 34–5, 115–6, 148
algebraic curve 130
altitude 76, 111, 137, 267
angle
 – adjacent 93–5
 – alternate 93–5
 – chord-tangent 103, 106
 – at circumference 14, 21, 102–4
 – corresponding 93–5
 – exterior 101
 – interior 90, 98
 – normal 107
 – opposite 93–5
 – reflex 90
angle bisector 60, 111, 125
ANGLE-BISECTOR macro 60
angle marking – see **Mark an angle**
angle sum 96, 99–102
Appearance of objects command – see **Look of objects**
arbelos 266
area 39–40
Aubel's theorem 83
axial affinity 200

Bernoulli's lemniscate 191
Brianchon's theorem 184

calculations 20

cardioid 190
centroid 17, 43, 120, 123–4, 264
Check property command 21, 38, 257
CHORD-TRISECTION macro 159
circle 11 *and throughout the book*
CIRCLE-C-r macro 66, 208
CIRCLE-SEGMENT macro 161
circumcentre 36, 45–6, 77, 166, 264, 267
circumcircle 43, 45–6, 76, 78, 103, 111, 113, 119, 123, 135, 165–6, 267, 269
cissoid 130–2
conchoid 134
Configurative Mobility Principle 36
conic 128–30
Construction menu 13–5, 26, 44, 58, 205–10
COPY-ANGLE macro 40, 69, 166, 208
COPY-LENGTH macro 40, 68, 166, 208
Creation menu 11–3, 26, 58, 205–7
customisation 21, 26–9, 58, 204–5
cyclic quadrilateral 103, 105, 182–3, 235
cycloid 140

Delete relations command – see **Redefine an object**
dependence 25–6
Diocles' cissoid 130–2

279

directrix 138, 192
dragging 12 *and throughout the book*
drag-mode 33–4 *and throughout book*

Edit menu 16–8, 58, 205
Edit menus command 27, 205, 222
ellipse 126–7, 129, 188–91
enlargement 35, 219, 247
ENLARGEMENT macro 249
EQUILATERAL-TRIANGLE macro 52, 146, 162–3
Euler line 264
Exposition command 29
EXTERNAL-TANGENTS-TO-CIRCLE macro 228–9

Fagnano's problem 79, 112
Fermat point 81, 113
File menu 205–6
focal point – see focus
focus 138, 188, 192
frieze patterns 150, 211

glide reflection 138, 187, 249
GLIDE-REFLECTION macro 249
Grid command 21

Hart's linkage 224
Help command 55
hexagon 116–8, 148, 185–6, 214–5
History command 21, 29, 51, 54, 219, 221–2
h macros 227–8
hyperbola 127, 130, 190–1
hyperbolic geometry 227–8

incentre 165–7
incircle 155–6, 164–7
INCIRCLE macro 156
INSERT-TRIANGLE macro 54
insertion problems 51–5, 113, 121, 152–3, 196-8
Intersection command 34, 176

Invariance Principle 36, 88, 109
inversion 125, 127, 194–6, 209–10
INVERSION macros 210
invertor – see Peaucellier and Hart
iso-areal problems 218
isoperimetric problems 217, 232–242

Johnson's theorem 265

kite 47, 85, 235
KITE-RHOMBUS macro 47
Klein's hyperbolic geometry 227–8

Label command – see **Name**
LAY-DOWN-QUADRILATERAL macro 213
lemniscate 191
limaçon 42, 133
line reflection 138, 245
LINE-REFLECTION macro 249
LINE-REFLECTED-TRIANGLE macro 211
LINE-SYMMETRIC-PT macro 66
line symmetry 142–151
Link a point command – see **Redefine an object**
loci – see locus
locus 23–5, 41–3, 52–3, 119–141, 149–50, 188–97, 238–9, 269
Locus command – see locus
LOGO language 9
Look of objects command 17, 26, 222

Maclaurin's trisectrix 132
Macro command – see Macro
Macro-construction – see Macro
Macro 9, 27–9, 43–9, 51, 57–71
Mark an angle command 19
Measure command 19
median 17, 33, 76, 111–2, 119, 264
menu reduction – see customisation
MIDPOINT macro 60
Miquel point 267

Miquel's theorem 266
Miscellaneous menu 18–22, 205–7
Morley's theorem 75

Name command 16
Napoleon–Barlotti theorem 39, 115–6
Napoleon's theorem 82
Newton's strophoid 132
Nine-point circle 135, 267
non-Euclidean geometry 227–8

orthocentre – see orthocircle
orthocircle 17, 24, 27–8, 75, 135–7, 264, 267

pantograph 126
Pappus' theorem 262–3, 266
parabola 24, 127, 130, 190, 192
PARALLEL macro 63
parallelogram 38–40, 84–5, 95–6, 109, 114–5, 147, 178–82, 196–7
parameter 40, 199
Pascal's limaçon 42, 133
Pascal's snail 190
Pascal's theorem 184, 189
Peaucellier's Cell 127, 194, 224
pedal curve 189–192
pedal quadrilateral 115
pedal triangle 79–81
pentagon 19, 49, 148
perimeter – minimal 79, 112
PERP-BISECTOR macro 59
PERP-CONSTRUCT macro 62
PERP-ERECT macro 61
perpendicular bisector 36, 59, 88, 111
perpendicular, construction of 61–3
Pictorialistic Hypothesis 35, 73
polygon (see also square etc) 100–2, 181, 212, 241–2, 268
PLC-EQTRIANGLE macro 162–3
projection 127
proof (see also **Check property**) 21, 85–6, 256–7
PT-SYMMETRIC-PT macro 64–5

Pythagoras' theorem 20, 43, 85, 263

quadrilateral 38–40, 46–8, 82–3, 99–100, 103, 114–6, 181–6, 196–7, 235–8, 253
QUAD-KITE macro 47
QUAD-SQUARE macro 47

reciprocating pump 225
Record command 21
rectangle 96, 103, 109, 147, 182, 235–8, 240
Redefine an object command 110, 119, 138, 172, 180, 185
reflection 64, 138, 187, 245, 249
REGULAR-HEXAGON macro 146
REGULAR–OCTAGON macro 146
REGULAR-PENTAGON macro 146
revolving table problem 222–4
rhombus 40, 47, 84, 147, 182, 235
RHOMBUS-SQUARE macro 47
Richmond's construction 49
rotation 35, 246
ROTATION macro 249

Scarab beetle curve 193
seven circles figure 142
SHAFT macro 226
simplification – see customisation
Simson line 267, 269
square 38, 46–8, 82–3, 99, 121, 147, 152–3
SQUARE-IN-TRIANGLE macro 154
SQUARE macro 38, 146, 153, 198
strophoid 132
subdivision of line segment 262
Symmetrical point command 143, 177, 207
symmetry 142–151

TANGENT-PT-CIRCLE macro 166
TANGENTS-TO-2-CIRCLES macro 210
tessellation 213

Thales' theorem 104
Thébault's theorem 38
touching circles problems 220–2, 265
trammel wheel 224
transformation 35, 126, 199–202, 243–9
translation 35, 244
TRANSLATION macro 249
trapezium 85, 115, 174–5, 235, 239
triangle 12 *and throughout book*
trisection of angle 75
trisectrix 24, 132

Van Shooten's theorem 78
Varignon's theorem 109, 114
Viviani's theorem 37, 77

Wallace's theorem 267
wallpaper patterns 151

3-POINT-CIRCLE macro 166, 235, 253
7-CIRCLES macro 143–4
7-CIRCLES-REFLECTED macro 143–4

Geometry with Cabri

This series of Cabri-Géomètre support materials begins with three co-authored by Chris Little and the series editor Rosamund Sutherland.

Taking a New Angle Chris Little and Rosamund Sutherland This booklet presents some ideas for using Cabri-Géomètre to explore elemetary properties of angles at a point, between parallel lines, and in triangles and polygons. The activities may be photocopied.

Contents: Introduction, Getting started with Cabri, Angles in Cabri, Angles around a point, Vertically opposite angles, Alternate angles, Angles in triangles, Angles in a polygon.

Activities: Introduction to Cabri-Géomètre, Constructing an angle, Angles around a point, Vertically opposite angles, Alternate angles, Corresponding angles, Finding alternate and corresponding angles, Opposite, alternate and corresponding angles, Angles in a triangle, Angles in polygons.
January 1995, 20 pages, ISBN 0-86238-377-3.

Exploring tigonometry Chris Little and Rosamund Sutherland
This booklet presents ideas and activities for using Cabri-Géomètre to teach trigonometry. The activities may be photocopied.

Contents: Introduction, Getting Started with Cabri (points,lines and circles, any equilateral triangle, angles on the circumference of a circle, any right-angled triangle) Trigonometry with Cabri (a paper construction for investigating SINE, constructing a triangle for exploring trigonometry, investigating SINE – the case of the 30 degree angle, is there a rule for any angle? introducing SINE of an angle, using the result of the investigation, calculating the SINE of an angle with a calculator, investigating the COSINE function, using COSINE and SINE)
A concluding note.

Activities: Introduction to Cabri-Géomètre, Any equilateral triangle, angles on the circumference of a circle subtended by a diameter, Any right-angled triangle, Constructing a triangle for exploring trigonometry, Investigating the SINE function – the case of the 30 degree angle, Investigating the SINE function – is there a rule for any angle? Using SINE, Investigating the COSINE function – the case of the 30 degree angle, Investigating the COSINE function – is there a rule for any angle? Making up trigonometry problems,Using COSINE and SINE.
January 1995, 26 pages, ISBN 0-86238-376-5.

Transforming Transformations
Chris Little and Rosamund Sutherland

This booklet presents ideas for using Cabri-Géomètre to analyse and explore the plane isometric transformations – reflection, translation, and rotation. Constructing images of plane figures using pencil and paper methods (for example using rules and compasses, or by means of counting squares on squared paper or graph paper) is time-consuming and not particularly satisfying. Once the transformations are constructed in Cabri, however, both the transformation itself and the object can easily be varied, and the basic properties can be investigated. We think that there is a real benefit in asking pupils to construct their own transformation macros since this provokes them to analyse the transformations for themselves. As they construct the macro, they make sense of the properties of the transformation. Each transformation is treated in the same way. First pupils build a macro which constructs the image of a single point. They then use this to construct a macro for the image of a triangle. With this macro, the basic properties of the transformation are explored, including combinations of transformations.

Contents: Introduction, Macro-constructions, Reflection (What is a reflection? Reflecting triangles, A puzzle) Rotation (What is a rotation? Rotating triangles), Translation, Two reflections.

Activities: Macro-constructions, Using symmetrical point, Creating a macro to reflect a point, Reflecting triangles, Reflection, Creating a macro to rotate a point, Investigating rotation, Rotating triangles, Rotation, Creating a macro for translation, Vectors, Two reflections.
January 1995, 34 pages, ISBN 0-86238-378-1.

Published by Chartwell-Bratt.

OTHER MATHEMATICS BOOKS FROM CHARTWELL-BRATT / STUDENTLITTERATUR
Watkins A J
DERIVE-Based Investigations for Post-16 Core Mathematics
Tony Watkins of the Centre for Teaching Mathematics, University of Plymouth, has produced these excellent practical exercises which show how to use Derive to discover mathematical concepts faster, have more fun in the process and achieve a deeper understanding. Developed over a 5 year period and extensively trialled, the investigations can be used by a wide range of students; from those with only high school mathematics to university undergraduates. The author's wide experience of using DERIVE as a highly effective teaching tool will provide stimulating and proven techniques for your own classroom.
102 pages, ISBN 0-86238-312-9, 2nd edition

Heugl, H., Kutzler, B
DERIVE in Education: Opportunities & Strategies
The main contributions from the 1993 Krems conference on DERIVE didactics are beautifully presented here. Experts from 14 countries examine a wide range of educational issues, offering suggestions and experiences at school, college and university level. This book will help you to teach and use maths in a faster, more efficient and more comprehensive way using DERIVE and looks at how maths teaching should be changed to take advantage of it.

Contents: Success and Failure in Mathematics – the effect of technology; Assessment of Mathematical Ability in the 'Light' of DERIVE; DERIVE in Mathematics of the 11th and 12th Grade; Some Methodical and Didactic Remarks on Examples for the Application of DERIVE in Teaching Mathematics at 'Gymnasium'; Number Theory with DERIVE – some suggestions for classroom teaching; DERIVE-centred Research at the University of Plymouth; Deterministische und stochastische Simulationen für die Schule; Learning Visually – DERIVE in the Calculus Laboratory; Extending Horizons using DERIVE to explore meaningful applications in calculus; The Austrian Research Project – Symbolic Computation Systems in the Classroom; Using DERIVE in the Third and Fourth Form of Grammar Schools in Austria;The 5th Class ACDCA Project in Austrian Grammar Schools;DERIVE for Students at the Age of 17 to 18; The Application of a Computer Algebra System as a Tool in College Algebra; Algebraic Reasoning and CAS – Freeing Students from Syntax?The Use of Graphics Calculators and Computer Algebra Systems – differences and similarities; Understanding and Reflective Abstraction – learning the concept of derivative in the computer environment;A Mathematical Model of a Firebreak using DERIVE; Using Computer Algebra Systems in Teaching Mathematical Modelling; Using DERIVE in Teaching Calculus Geometric Maximum and Minimum Problems; Using Computers – An Experience in Algebra and Discrete Mathematics; The World System of Johannes Kepler in Stereo-Vision with DERIVE; From Harmony to Chaos; Using DERIVE in the Calculus Classroom – a step towards the future.
302 pages, ISBN 0-86238-351-X, 1994

Ed. Josef Bohm
Teaching Mathematics with DERIVE
This illustrated guide is packed full of sound advice and teaching materials. Top European educationalists show how to use computer algebra systems to teach mathematics (particularly to 12–18 year olds). It is full of helpful suggestions for teachers and shows the impact on different parts of the curriculum. If you are teaching mathematics, at school, college, or university, these papers from the 1992 Krems conference will be useful and thought-provoking.
298 pages, ISBN 0-86238-319-6

Sjostrand, David
Mathematics with Excel
Leads the reader through some beautiful parts of the mathematical landscape showing that Excel and Derive are superb tools for exploring mathematics. Shows how Excel can be used to visualize important mathematical ideas and make them concrete. The graphical features of Excel are used frequently.

Intended for students at secondary school and college level. It contains many examples with detailed instructions and exercises. It can be used by students with little or no experience of spreadsheets as well as by advanced Excel-users. The reader is assumed to have access to Excel 5.0 although readers with older versions or other spreadsheets will find inspiration in the book.

Contents: Worksheet formulas; Copying formulas using absolute and relative references; Customizing Excel using Visual Basic macros and user-defined functions; Graph plotting in 2 and 3 dimensions; Shifts, reflections and rotations of curves; Numerical calculation of derivatives and integrals; Numerical solution of equations; Systems of equations and differential equations; Direction fields; Simulation of motion; Matrices; Linear and nonlinear optimization; Sets and algebraic structures; Groups; Combined use of spreadsheets and computer algebra (DERIVE).
190 pages, ISBN 0-86238-361-7, 1994

Andrew Rothery
Modelling with Spreadsheets
"Modelling with Spreadsheets" explains and illustrates both the principles of modelling and the use of computer spreadsheet methods. It describes several well-known applications of modelling in detail and provides spreadsheet activities for the reader to tackle on his or her own computer. A major theme of the whole book is the modelling process: 'optimisation'. The applications are mainly in the business/economics area and illustrate how modelling can be used to inform decision-makers.

The only mathematical prerequisite the reader needs is an ability to cope with calculation rules and simple formulas. The theoretical principles underlying the processes and strategies of modelling and optimisation are fully explained at an introductory level.

"Modelling with Spreadsheets" is clearly written and accessible. The general reader, teacher and business spreadsheet user interested in spreadsheet modelling will find it enjoyable and informative. The student reader, for whom the book is ideal, will be following a course in Computing, Business, Economics, or study of modelling or spreadsheet techniques in a business/economics context.
63 pages, ISBN 0-86238-258-0

Harry V Smith
Numerical Methods of Integration
Describes, with aid of worked examples and supplementary problems, many recent and important techniques for numerical evaluation of definite integrals. Nine chapters cover Newton-Cotes, Gaussian, Kronrod and SINC quadratures. Methods are included for numerical evaluation of integrals whose integrands contain singularities, in particular Cauchy principal value integrals, and for numerical evaluation of divergent integrals, as well as for approximation of infinite and semi-infinite integrals.
Also presents sets of tables of global error bounds for most important quadrature rules; letting you choose which form of a quadrature rule will approximate an integral to any accuracy required in most applications after only one calculation.
147 pp, ISBN 0-86238-331-5, 1993

Lennart Rade and Bertil Westergren
Beta Maths Handbook
BETA is a comprehensive mathematics handbook for scientists, engineers and university students. It presents, in a lucid and accessible form, classical areas of mathematics such as algebra, geometry and analysis. Furthermore, areas of particular current interest are covered, such as discrete mathematics, probability and statistics, optimization, programming and numerical analysis. BETA concentrates on definitions, results, formulas and tables. The importance of computers in modern applied mathematics is also considered. If you are an advanced mathematics user this excellent
reference source will make life easier.
425 pp, ISBN 0-86238-140-1, Hardback

Zhe-xian Wan
Introduction to Abstract and Linear Algebra
That linear algebra is an indispensable tool in engineering science is well-known. Since the mid of this century abstract algebra as well has found more and more applications within the area. For instance, finite fields play a prominent role in coding theory, and ring theory is of course the foundation of linear systems over rings.
Both linear and abstract algebra should now be in the curriculum of undergraduate engineering students. This introductory book on algebra provides the basic material for such a course. It also constitutes a solid algebraic basis for the non-specialists who wish to become specialists in, e.g., coding theory, cryptography, and linear systems theory.
The important algebraic structures, such as groups, rings, and fields, are introduced and the main contents of linear algebra are included. Many examples are given to illustrate the abstract concepts and the computational methods are emphasized. Finite fields and the canonical forms of the various kinds of matrices over finite

fields are treated in some detail. The book concludes with a chapter on polynomials over finite fields, which utilizes both the abstract and the linear algebra studied in the previous chapters.

The only prerequisite needed to read the book is high school mathematics, and it is also suited for self-tuition. Exercises are given at the end of each chapter.

Zhe-xian Wan
Geometry of Classical Groups over Finite Fields
This monograph is a comprehensive survey of the geometry of classical groups over finite fields. Many important results, appearing for the first time in book form, are clearly presented. Most of them have been obtained by the author and students under his supervision.

The book is written in a leisurely style and will be a good reference for those who work on classical groups, finite fields, finite geometry, combinatorics, block designs, and coding theory. To read the book only basic knowledge of abstract and linear algebra is required. Exercises are given at the end of each chapter.

The first two chapters on affine geometry and projective geometry are rather classical subjects, and they are very thoroughly presented. The chapters also contain new results and new derivations of old results. Chapter 3 to Chapter 7, the main contents of the book, treat the geometry of symplectic, pseudo-symplectic, unitary, and orthogonal groups over finite fields. Connections with polar spaces and generalized quadrangles are pointed out, and applications to error-correcting codes, authentication codes and projective codes are given as examples or exercises.

Figures and Macros

The Cabri figures and macros to accompany this book are available from Chartwell-Bratt on 3.5" disks. Please state whether you require MS-DOS or Macintosh formats and enclose £3 (plus vat within EC). VISA and Mastercard welcome.

Alternatively get them for free on the Internet

Chartwell-Bratt Ltd, Old Orchard, Bickley Road, Bromley, Kent, BR1 2NE, England, tel (+44) (0)81 467 1956, fax (+44) (0)81 467 1754
Internet philip@chartwel.demon.co.uk